The Universe Natural History Series

THE LIFE OF PRIMATES

The Universe Natural History Series

Editor: Richard Carrington
Associate Editors:
Dr. L. Harrison Matthews
Professor J. Z. Young

The Life of Primates

ADOLPH H. SCHULTZ

UNIVERSE BOOKS

NEW YORK

Published in the United States of America in 1969
by UNIVERSE BOOKS
381 Park Avenue South, New York, N.Y. 10016

Copyright © 1969 by Adolph H. Schultz

Second printing 1972

Library of Congress Catalog Card Number: LC 75–91990
SBN 87663-108-1

Printed in Great Britain

Contents

List of Plates

Acknowledgements

The author and publishers wish to thank the following for providing photographs for this volume: University of Chicago Press, plate 17; Zoological Society of London, plates 18, 19, 24, 25, 26, 28, 29, 34, 35; Dr B. Grzimek and Frankfurt Zoo, plate 20; J. Klages, plates 21, 32; National Geographic Society, plate 36; Beringer and Pampluchi and Zurich Zoo, plate 22; Antwerp Zoo, plate 23; 20th Century-Fox, plate 27; E. P. Walker, plate 30; Natural History Zoo, Berne, plate 31; *Das Tier*, plate 33; da Cruz Lima, Eladio, plates 37, 38, 39.

Preface

IN BOTH their body build and behaviour monkeys and apes are unquestionably the most man-like of all mammals and have therefore aroused special interest since ancient times. For public entertainment they form the most popular attraction in our zoos, they have played many roles in legends and mythology and have figured in literature and art with remarkable frequency. In recent years our nearest animal relations have come to be widely used in countless scientific experiments for which no other animals could serve as advantageously. Due to the close physiological similarity between apes and man the former were chosen as pioneers in orbital flight in preparation for the exploration of space and much of the progress in modern medical and psychological research could not have been gained without our simian cousins as servants of science. Ever since Linnaeus had placed man in the mammalian order of primates with lemurs and monkeys and, particularly, since Darwin's epochal claim of a common origin for all primates, they have been studied with steadily increasing intensity by more and more specialists. There exists already a large international society of primatologists and two new journals are devoted exclusively to this young science. In 1941 there appeared a most useful classified bibliography of chiefly the morphological and physiological papers on nonhuman primates, containing 4,630 titles (Ruch, ed.). The later publications, dealing with far more aspects of primatology, have been overflowing into a vast and widely scattered literature which requires heroic efforts to collect and today merely the titles could easily fill many volumes. Large parts of this literature are currently being reviewed and condensed by a surprising number of writers interested in limited aspects of the entire flourishing field of modern primatology. Outstanding among the most recent surveys are the serial volumes by Osman Hill, devoted to a systematic enumeration of all living and fossil primates, species by species, and the handbook of primatology, edited by H.Hofer, D.Starck and the writer with the co-operation of many specialists, which deals mainly with general morphological topics. Though many volumes have already been issued of both these comprehensive technical works, they have not yet been completed. Most of the many further and shorter reviews or collections of significant papers have been prepared for limited purposes, such as accounts of primate evolution with emphasis on fossil finds, surveys of the social behaviour of monkeys and apes, or the frequent attempts to describe the main characters of higher primates as a basis for a more detailed discussion of human evolution. Some prominent examples of this profusion of more or less comprehensive, generalising publications on primates have been listed in the bibliography at the end of this book under the first sub-heading. Additional titles of such reviews of special aspects of primatology will be found in the lists of selected literature appertaining to other chapters.

In view of this already very extensive supply of technical and popular accounts of primates in general with varying aims and limits it may seem to be quite superfluous to present one more volume devoted to this same mammalian order. The only justification, which the writer can offer for having accepted the invitation to contribute this book, is his belief that his own interests, developed during five decades of work on primates, may provide a somewhat different focus from that of other primatological authors by emphasising and interpreting certain aspects of the life of primates in the light of his own experience. Being primarily a physical anthropologist by training and academic position, the writer, naturally, has paid most attention to such comparative studies which are essential for our understanding of the primate nature of man. This has led to a somewhat more detailed consideration of the simian primates than of the prosimian ones. If systematics, behaviour and many other important topics have been dealt with too briefly in the opinion of the respective specialists, it can merely be pointed out that modern primatology has expanded to such a vast field that no attempt to review it can cover more than selected interests without becoming not only very technical, but also far too voluminous. The actual selection of the major subjects to be discussed is readily seen from the chapter headings, but it will be discovered only later that these titles merely indicate the central interest of the respective discussions in which some repetition or overlapping could not be avoided. For instance, secondary sex differences are dealt with not only in a separate chapter, but had to be referred to also in describing development, behaviour, dentition, etc.

All the pen-and-ink and half-tone drawings in this book have been made by the author himself with the aid of specimens or photographs in his collections. Many have been prepared specially for the present purposes and others had already been used for previous publications by the author and were merely retouched. The places of former publication of the latter drawings can be found from the titles of the figures and the bibliography. For the generous consent to reproduce the various photographs in this volume the author is very grateful to many of his colleagues and their publishers, as listed under Acknowledgments. Last but not least, the writer wishes to express his sincere thanks to his friend, Professor J. Biegert, his successor in charge of the Anthropological Institute of the University of Zürich, for having provided a comfortable room in this institute, where the author has had convenient access to the institute's and his own collections of material and literature while preparing this book.

Chapter 1

Historical Introduction

MAN'S fascination with the animals most similar to himself dates back to antiquity. The ancient Egyptians venerated baboons, as is attested by representations on their numerous monuments and by carefully embalmed bodies of these primates. Several other species of African monkeys, particularly guenons, have been found engraved or painted on the walls of early tombs and temples. The old Babylonians, Hebrews and Etruscans also seem to have been well acquainted with various kinds of monkeys, which they must have obtained as ambassadorial presents or in trade. Aristotle had summarised the astonishingly wide knowledge of pre-Christian Greece regarding tailed and tailless monkeys, described under five different names, by emphasising their resemblance to man. Old World monkeys have played parts in Greek and Roman mythology and were then commonly regarded as grotesque caricatures of man, pretenders to human status or as symbols of ugliness. To orthodox Hindus the large langurs of India are sacred, being identified with their legendary god Hanuman, who with his monkey followers had aided Rama in his battle with the king of demons, Ravana. Antedating even the famous Ramayana, statues of Hanuman were already used as guardian gods of every settlement and to this day all monkeys are at least tolerated in many parts of India, with the result that they often become impudent pests. The macaques of China and Japan have been favourite subjects for admirable representation in painting and ivory carving for many centuries, often being shown in action with other animals or man himself as illustration of legends or fables, but not as creatures to be ridiculed, feared or despised, as commonly seen in Western art (plate 1). In other parts of the world, inhabited by nonhuman primates, man's attitude toward them varies between opposite extremes. While some species are never molested or are even held in superstitious awe, others are unhesitatingly hunted for their meat or hides, like any other game, or killed when raiding the

I

natives' plantations. Orang-utans have gained their Malay name of 'Man of the Woods' from the formerly widespread belief that they are merely a hairy variety of human beings, who could even talk, but prefer not to. Many nocturnal prosimians were regarded by the natives of Madagascar as reincarnations of deceased men and hence to be strictly avoided. In some of the islands with tarsiers on them these small ghost-like creatures are never referred to by their name for fear of becoming haunted. In Western Europe during the middle ages and the renaissance African and Asiatic monkeys as well as apes were used with surprising regularity in folklore, art and literature in the roles of fools, imitators, tricksters and sinners, or as representing vanity, lust and still other abhorrent traits. With increasing frequency they were chosen as human caricatures for satirical writing and drawing in critiques of social conditions or artistic fashions (plate 3). For many centuries monkeys trained to stand upright and to wear human clothing were favourite subjects for entertainment at courts as well as for the amusement of the people (plate 4).

Reports of ancient explorers and returned sailors concerning large, hairy, tailless, upright creatures in the caves and jungles of foreign countries, mostly based on mere hearsay and embellished with every repetition, evidently referred to one or another of the man-like apes and may have originated, or at least endorsed, the common belief in the existence of fauns, satyrs, sphinxes and other fabled beings, resembling man. The best known of such early stories is that of the Carthaginian admiral Hanno, who in the fifth century BC had returned from a voyage to the west coast of Africa with the hairy skins of three 'wild women', called 'gorillas', which his men had killed after savage resistance by the strange creatures. In the faithful translation by Falconer (London, 1797) of the Greek version of the original, but lost, Carthaginian report of the *Voyage of Hanno* this puzzling encounter is described in part as follows: the party came to 'an island full of savage people, the greater part of whom were women, whose bodies were hairy, and whom our interpreters called Gorillae . . . Three women were taken . . . and could not be prevailed upon to accompany us. Having killed them, we flayed them and brought their skins with us to Carthage.' Since Hanno did not get nearly as far south as the present range of gorillas, it seems more likely that he had encountered chimpanzees, but in either case he had supplied some tangible evidence for the popular belief in beings which seemed to be neither man nor animal.

A more rational approach to an understanding of the true nature of nonhuman primates had its vague beginning in the second century AD when the famous Greek physician Galen had not only dissected, but even experimented with the common stumptailed African macaque, the so-called Barbary ape, for his influential anatomical investigations. Galen, however, had simply taken it for granted that the bodies of apes resemble human anatomy so closely that they can be advantageously substituted for inaccessible human cadavers. The inaccurate results of this assumption remained generally uncorrected until the sixteenth century, when the great Belgian anatomist Vesalius dared to attack the long prevailing Galenist teaching and thereby stimulated further study of physical differences between man and apes. Under this influence the British surgeon Tyson undertook the first and still exemplary dissection of a young chimpanzee, then called 'Orang-Outang, sive Homo Sylvestris'. This he published in 1699 in a splendid monograph with large plates, illustrating many anatomical structures, and with detailed comparative descriptions in the text, enumerating all the conditions in which his rare specimen resembled man more closely than do monkeys. In the lengthy title of this classic volume the author's ultimate aim and chief conclusion is clearly stated, namely that there is added 'A philological essay concerning the Pygmies, the Cynocephali, the Satyrs, and Sphinges of the Ancients. Wherein it will appear that they are all either Apes or Monkeys, and not Men, as formerly pretended.' Even though Tyson's widely-read treatise contains no word regarding real relationships between the primates under comparison, it was nevertheless very helpful in preparing the scholarly world for the reception of the idea of evolution in general and of that of man in particular. The admirable drawing of Tyson's young chimpanzee in a fully upright pose, aided by a cane in its hand, made such a deep impression that apes were commonly shown as holding themselves erect with some stick or pole for support (plate 2). The skeleton of Tyson's 'Pygmy' is now in the British Museum (Natural History).

Ever since Marco Polo, more or less fanciful accounts of travellers regarding nonhuman primates accumulated and were collected by the numerous writers of early natural histories without significantly advancing science, but revealing the credulity and superstition of the time. Not until the middle of the eighteenth century was a serious attempt made to bring some order in these seemingly endless

3

lists of old and new descriptions of animals by grouping them according to their shared qualities. This was accomplished by the brilliant Swedish botanist Linnaeus, who introduced the binominal nomenclature in his famous catalogue, the *Systema Naturae*, which in its tenth edition of 1758 formed the starting point of modern classification and began with a first order, named *Primates* or primary animals, having man in the first place. The creatures assigned to this order were selected as having in common a combination of the same characters which distinguishes them from all other mammals. His *Primates* Linnaeus defined as possessing four parallel upper teeth of the 'first kind' (= incisors), two mammary glands on the chest, 'palms which are hands' and fully formed clavicles, and this order he subdivided into four genera, namely *Homo, Simia, Lemur* and *Vespertilio* (= bats). In the first edition of the *Systema* in 1735 the order was still called *Anthropomorpha* (= of human form) and included only *Homo* and *Simia*, besides *Bradypus* (= sloth). Under *Homo* are listed the species *sapiens* with several races as well as a second one, called *Nocturnus*, represented by some of those fabled hairy creatures of earlier authors under specific names of *Troglodytes* or *Sylvestris* (= cave- or forest-dwellers) and supposedly occurring in Ethiopia and Java. In his genus *Simia* Linnaeus placed all monkeys and apes of the Old and New World then definitely known and his genus *Lemur* contained not only the few prosimians, described so far, but also the 'flying lemur' *Galeopithecus* which actually belongs to a separate order of Dermoptera.

The daring but innocent inclusion of *Homo sapiens* with apes and monkeys in one and the same order of animals by a famous and really very pious author appeared in those days as such a shock to the dignity of man that new classifications were proposed in rapid succession, stressing supposedly fundamental human distinctions. In a lavishly illustrated set of volumes on mammals, which appeared in 1775, the German zoologist Schreber was the first to emphasise the physical uniqueness of man among primates by having only two hands, instead of four. This supposedly radical difference was endorsed by one of the great pioneers in anthropology, Blumenbach, who in 1791 coined the formal names of *Bimanus* and *Quadrumana* for two separate orders, the former being reserved for man alone. He insisted that the feet of apes, monkeys and lemurs are generally even better suited for grasping than their hands, whereby they are clearly separated from other quadrupedal mammals and from bipedal man.

4

This separate classification of man was accepted by the great and influential French zoologist Cuvier and persisted in some places even to the end of the last century. Professor Illiger, a German zoologist, had in 1811 unsuccessfully introduced the name *Erecta* for an order of man alone, maintaining quite rightly that the permanent erect posture constitutes a more significant distinction than the possession of only two hands. The contemporary compatriot of Darwin, the eminent comparative anatomist Sir Richard Owen, attested also his firm belief in the profound uniqueness of mankind by assigning it in 1868 not only to a separate order, but even to a special subclass of the animal kingdom – the *Archencephala* or possessors of superior brains. These and many other major and minor variations in the classification of primates during the century between Linnaeus and Darwin were supported by a steadily increasing mass of publications devoted to the natural history and comparative anatomy of the recent species. Such ambitious monographs as the huge and lavishly illustrated *Histoire naturelle des Singes et des Makis* by Audebert (Paris, 1799) or the boastfully entitled *Die vollständigste Naturgeschichte der Affen* by Hofrath Reichenbach (Dresden, 1863) with its 501 coloured figures testify to the western world's intense interest in primates and reveal the emancipation from superstition, the diminishing influence of theological and philosophical preconceptions, as well as a rising concentration on direct observation of the living and dead animal.

By far the greatest and most productive impetus for the study of primates came with the publication of the epochal work on evolution by Darwin and particularly with that challenging paper of 1863 by the aggressive Thomas Huxley, entitled *Evidence as to Man's Place in Nature*. Mere classifying according to similarity had now gained new meaning by implying common origins, while distinctive characters could be explained as adaptations, evolved through selections, and taxonomic systems could be interpreted as family-trees. Comparative anatomy and embryology flourished as never before and became triumphantly supported and supplemented by rapidly increasing numbers of newly discovered fossil remains, which provided the most convincing evidence for evolutionary change. Though the theory of evolution was born in England, it thrived at first most lustily in Germany with its enthusiastic advocate E. Haeckel and then particularly under the amazingly productive leadership of the Würzburg anatomist Gegenbaur and his

5

entire school of collaborators and successors, who contributed a steady stream of papers of lasting value on primate anatomy.

No attempt can here be made to outline the involved history of this great post-Darwinian progress in primatology. It must suffice to mention that scientific controversies did not disappear, but rather multiplied with the discovery of new facts, open for new interpretation, and with the new problems of the exact phylogenetic interrelations between the different groups of recent and extinct primates. Indeed, many of these problems have not yet been solved with general agreement to this day, especially those concerned with the proper assignment of some fossil forms, available in tantalising fragments only, or with certain basic questions regarding the ages and places of evolutionary divergence of some primates and their most appropriate classification. For instance, while comparative anatomists had soon been able to demonstrate that such a one-sidedly specialised form as the aye-aye of Madagascar is a primate and not a rodent, as formerly believed on account of its curious front teeth, scholars are still debating whether the tree-shrews should be placed among the insectivores or the primates according to the differing significance ascribed to some of their characteristics. There is general consent among palaeontologists of today that the famous skull cap of *Pithecanthropus*, found in Java toward the close of the last century, belongs to an early form of man and is not that of a giant gibbon or prehistoric chimpanzee, as had once been claimed. On the other hand, even some of the foremost experts still hold quite different views in regard to the proper systematic position of the exceptionally complete fossil remains of *Oreopithecus*, an ancient ape from Italy. Discoveries of new types of not only extinct, but even of living primates seemed endless during the post-Darwinian century and continued to fill the major museums of the world. The largest of all recent primates, the gorilla, remained hidden from the civilised world until 1847, when the missionary T. S. Savage and the Harvard anatomist J. Wyman had published a 'Notice of the external characters and habits of *Troglodytes Gorilla*, a new species of orang from the Gabbon river. – Osteology of the same' (*Boston J. of Nat. Hist.*, 5). The somewhat different eastern or so-called mountain gorilla did not become known even until the beginning of our century, when it was briefly described in 1903 from a large male specimen, shot the year before by a German officer, named Beringe, after whom the new race is named. A hitherto unknown, small species of chimpanzee from the left side of the Congo river was discovered in 1929 and a rare marmoset from the Amazon region with an exceptional dental formula in 1904. The fossilised skull of a young primate, resembling chimpanzees, was found in South Africa in 1925 and received the name *Australopithecus,* or southern ape. This was not generally acknowledged as a close relative of man until many further remains of the same genus left no doubt as to the true nature of this highly important discovery. Surprisingly large fossil teeth of evidently another new form of a higher primate were found shortly before the last world war in Chinese drugstores of Hong Kong and Canton among the 'dragon teeth' for sale as supposedly rejuvenating medicine. They were at first regarded as belonging to some huge precursor of man and christened *Gigantopithecus,* but later finds of several entire mandibles with the same kind of teeth showed that these precious grinders came from big apes which roamed in China about half a million years ago. These random examples of important additions to our newer knowledge of primates could easily be extended into such a long list that it is not surprising that there also occurred a few premature rejoicings which did not survive later critical studies. Some of the latter disappointments were of special interest due to the fact that no recent or fossil species had ever been found in the New World which could claim any close relation to that distinguished group of higher types, among which the progenitors of man are to be expected. For instance, at the start of this century the Argentinian palaeontologist F. Ameghino had insisted that certain fragmentary fossils, found by his brother in Patagonia, represented missing links between the simian primates of the Old and the New Worlds and even an early, small forerunner of man, which he persuasively named *Homunculus.* Unfortunately later study revealed this creature as merely a primitive, but typical American monkey. In 1922 the eminent palaeontologist H. F. Osborn had created a sensation with the announcement of the discovery of a fossil tooth of a supposedly man-like ape in North America, a true 'western ape', published under the proud title '*Hesperopithecus,* the first anthropoid primate found in America'. From subsequent investigations of further material from the same locality, however, it appeared that this specimen is a badly worn molar of most likely an early type of peccary and not a primate at all. The hope of finding even living new apes, surviving in remote jungles of America or elsewhere, has never been given up. Thus, the credulous French anthropologist G. Montandon startled his colleagues in 1929 with a perfectly serious and much discussed report of apes in Venezuela, over five feet tall, which had attacked a geologist, who had been able to save only a photograph of the dead animal, posed on a box, plus the assertion that it lacked a tail. The picture, however, leaves no doubt that this creature, which had already been given the new generic name of *Amer-Anthropoides,* shows in every detail a typical spider monkey, whose tail was either shot off or else is hidden behind the box, whose actual size was unknown. Rumours of still other strange apes living in various tropical forests seem to be ineradicable, but so far have been no more convincing than the story of the Himalayan snowman, the fabulous *Yeti.* It seems a great deal more likely that new species of past ages will continue to be added to the long list of already known primates. The proper classification of the latter has greatly improved in recent years and our understanding of their evolutionary histories and interrelationships has advanced not only through the intense work of palaeontologists and comparative anatomists, but also through the modern means for determining the degrees of similarity in features which it had previously not been possible to take into consideration. With the discovery of exact methods for investigating the number and detailed shape of chromosomes and the intricate biochemical qualities of the blood there were gained very promising methods for advancing our knowledge of the genetic relations between the living primates, with the ultimate aim of determining man's exact position among them. The spectacular recent improvements in neuro-surgical techniques, combined with some electronic inventions, have opened new paths for exploring the actions of and reactions to the functioning brain. For such experiments the nonhuman primates are naturally the most suitable animals and have already served in highly significant studies of some phases of behaviour. Experimental work on many other physiological problems, such as menstruation, reproduction, manual dexterity, etc. has also been carried out most profitably on monkeys and apes and is currently being much expanded. In few other fields of science has there developed such a sudden and widespread interest as in the modern investigation of primate behaviour in general and of their social relations in particular. Started by such eminent pioneers as the late Professor R. Yerkes with his large group of collaborators at his famous chimpanzee laboratory in Florida, the brilliant Sir Solly Zuckerman with his stimulating survey of primate social life, and Professor R. Carpenter in exemplary monographs on wild howler monkeys and gibbons in particular, intensive work on primate behaviour is now pursued by a host of younger scientists in the field as well as in very many laboratory colonies. In view of the fact that man is the only primate able to speak, the means of communication of monkeys and apes, particularly the vocal ones, have been of outstanding interest for a long time and continue to challenge investigators. The famous Dutch anatomist Peter Camper had already in the eighteenth century dissected the larynx of orang-utans to find what he thought to be the reason why they could not talk like man. Since then many other attempts have been made to explain the profound human distinction of speech on anatomical grounds, even while the old belief had not been abandoned that apes could at least be taught to talk. For instance, at the close of the nineteenth century a remarkable American, named Garner, had devoted most of his life to propound his firm faith in 'monkey speech' in several entertainingly written books based largely on his extensive field work in African jungles. For the latter he had taken along a large folding cage in which he enclosed *himself* for studying in safety the voices of wild apes as they came to investigate him! Garner had evidently been impressed by the sensational reports of the French-American explorer Du Chaillu, who in 1861 had published lurid accounts of the extreme ferociousness of the gorillas he had hunted. Only in recent years has this slanderous characterisation of the largest man-like ape been fully corrected by the patient observations of the young behaviourist Schaller in his valuable report on the peaceful life of the East African gorilla.

This very sketchy account of the history of man's interest in his nearest animal relations to the rise of modern primatological research must suffice to introduce the following chapters which will deal with some of the most significant present results of the ceaseless investigation of the fascinating life of the primates.

Chapter 2

Introductory Survey of Primates According to their Main Distinguishing Features

AFTER having dealt with some of the very slow initial and rapid late advances in our knowledge of primates during past centuries, and before discussing single aspects of modern primatology it seems highly desirable to introduce this mammalian order of primates by defining its general nature and by describing briefly the distinguishing characters of its main subdivisions. It is also unavoidable to give here at least the most frequently used scientific names of the various classes, since fossils have no common names and many living forms have been given 'popular' names which are as unfamiliar to most laymen as the technical ones, besides being often just as long (e.g., Angwantibo for *Arctocebus* or Douroucouli for *Aotus*). This introductory survey will not only spare the writer endless repetitions of definitions in the later chapters, but will also enable the reader to appreciate the manifold evolutionary specialisations, acquired by different species or whole groups of them, by having learned their approximate places on the family-tree of the entire order, as they are indicated by their classification. The latter has been based chiefly and very properly on comparative anatomical investigations, which reveal the genetic relationship by means of degrees of morphological similarity. In the case of fossils this is naturally limited to similarities in bones and teeth. Even though all the enormous amount of work on recent primates is gradually being checked as well as enriched by comparative physiological and biochemical studies, as well as by the relevant findings of parasitology, by detailed descriptions of the chromosomes and by still other specialised means, no general agreement has been reached as yet in regard to all the problems of primate systematics. The following presentation, therefore, can merely give the currently prevailing views of the major interrelations and most essential characteristics of primates but for many details the reader may quite likely encounter differing versions of classification in the literature and, certainly, some other names for certain species or entire groups.

Well-stocked monkey houses of our larger zoos usually exhibit a great variety of types which appear to differ from one another in very impressive ways. Underneath their skins, however, most primates are really remarkably uniform and conservative when compared with the majority of other living mammals, which contain far more extreme and diversified specialisations within some single orders. The distinguishing morphological features of the primates have generally remained very modest in contrast to the innumerable one-sided adaptations which have profoundly changed so many mammals into swimming, flying and digging forms or into those outstanding in regard to bulk, speed, highly restricted diet, etc. This comparative lack of specialisation is the chief reason that no easy and clear-cut definition of primates can be given and that their few distinctive features are often rather unimportant ones, or have only general, but not constant validity. With these limitations in mind it can be stated that primates have first of all been adapted to a chiefly arboreal mode of life and in connection therewith have maintained a decidedly generalised limb construction, preserving particularly the primitive pentadactyl condition of five digits in most, but not all species. The first digits of the feet and the hands serve as a rule for grasping, but this function differs widely in its perfection. A variety of the lower species bear claws on at least some digits, while the great majority of primates possess flattened nails on all fingers and toes. All primates have retained fully functioning clavicles, which have become reduced or even entirely lost in so many other mammals. The dentition of primates has generally remained decidedly unspecialised in comparison with the profound changes in the teeth of a great variety of other mammals. The grinding teeth in particular have preserved a primitive cusp pattern and the trend toward reduction in dental elements, though very evident, has not progressed far, except in the curious aye-aye.

Also typical for primates are the clearly recognisable general tendencies to reduce the apparatus for smelling and to perfect at the same time the visual organs, especially the power of binocular and colour vision. Undoubtedly of greatest significance has been the progressive expansion and elaboration of the cerebral cortex and of the brain in general which must have begun to play its decisive role early in the evolution of primates as a distinct mammalian order.

The basic primate stock had become segregated during the ancient Palaeocene period, by which time it was already represented by several differentiated types. In the subsequent evolutionary history of primates the tempo of change was evidently very uneven, since entire groups have survived to this day which retain most of their original features with little alteration, whereas other and later forms have rapidly acquired advanced characteristics and comparatively few have progressed very far beyond the rest in many respects. For these reasons the different types of not only extinct, but also recent, primates clearly represent different stages along their common evolutionary pathways, a fact which greatly facilitates their classification.

According to the presently prevailing, but not universal, usage primates are divided into two suborders, namely the *Prosimiae* and the *Simiae*. The latter are also called *Anthropoidea* by some authors, but this designation causes much confusion since only the man-like apes are commonly called 'anthropoids' and also because the Greek term *anthropos* for man is hardly justified for such simian primates as marmosets. In Darwin's time the great Thomas Huxley had called only the higher primates *Anthropomorpha* to emphasise their man-like appearance and to distinguish them from the Old World monkeys which he called *Cynomorpha* after the dog-like baboons.

Of all recent prosimians the lively lemurs are most frequently shown in captivity and are hence the best known type of what in German is appropriately referred to as 'Halbaffen'. Though many prosimians do resemble monkeys in a superficial way, they have all remained definitely more primitive in their general construction and mode of life, in spite of the fact that prosimians flourished before any genuine simians had made their appearance. In view of their very long presence on earth and their formerly wide distribution in both hemispheres, it is not surprising that the suborder of prosimians has become highly diversified in general appearance as well as in regard to many anatomically localised specialisations. In body size they range between forms no bigger than a mouse to some, now extinct, species as large as a big male gorilla. A great many different prosimians are active only at night and have acquired large eyes, reaching a grotesque extreme in the small nocturnal tarsier. All prosimians mature more rapidly than their simian relations and quite a few species usually produce several young at a birth. Even

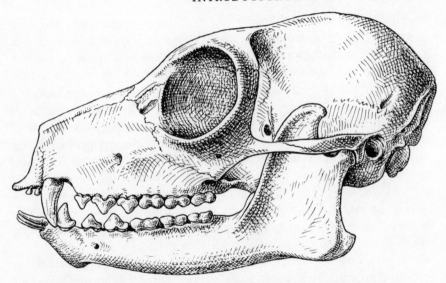

Figure 1   Side view of a typical skull of an adult lemur (*Lemur albifrons*).

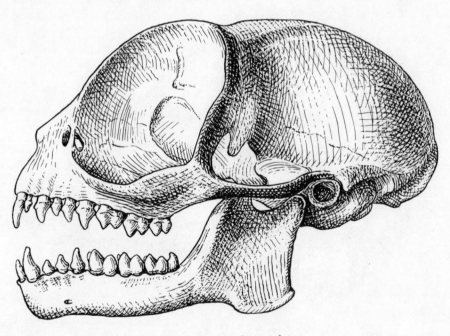

Figure 2   Skull of an adult female *Tarsius fraterculus*.

species with single offspring can still have more than the one pectoral pair of mammary glands along the ventral side of the trunk, and in that aberrant prosimian, the aye-aye, these glands are limited to the groins. The primitive condition of a forked or bicornuate uterus is typical for the suborder and so is the lack of a bony partition between the orbit and temporal fossa, which has appeared in an incomplete state only in tarsiers (figures 1 and 2). Many prosimians still have a long face with the snout projecting far forward as in most insectivores, but in other forms the jaws and entire face have become considerably reduced and even shifted toward the skull base, thus foreshadowing the typical conditions of simian primates. In a similar way the position of the eyes and their sockets varies widely among prosimians, some species still having them far apart and directed laterally, whereas in others they have approached each other much more closely without, however, being directed straight forward as in all simian primates (figures 3 and 4). An almost universal characteristic of prosimians can be readily seen on the outer face, where the nostrils are surrounded by hairless, glandular and moist skin, extending medially to the mouth. Such a so-called rhinarium exists also in many other mammals, but has

Figure 3   Head of an adult treeshrew (from a photograph of a living animal).

Figure 4   Head of an adult female *Lemur catta,* showing the typical rhinarium and vibrissae.

disappeared in the tarsiers and in all simian primates. As a further example of the striking interspecific and intergeneric variability of single features among prosimians it may be mentioned that the great majority of species possess long tails, but a few forms have become practically tailless. Similarly, most species have remarkably large and mobile outer ears, whereas in a few others the ears are so small as to be nearly hidden in the fur. In regard to locomotion some prosimians are slow and deliberate quadrupedal climbers while a great variety of others are distinguished by extreme agility and the power to jump incredible distances, propelled by their hindlegs alone. On the basis of these and many other deep-seated and technical details, prosimians have been subdivided into quite clearly distinguishable infraorders.

The first of these, called Tupaiiformes, is today represented by the treeshrews with a number of different genera. As mentioned before, not all authorities are willing to regard these creatures as genuine primates, since they have retained much resemblance to primitive insectivores. It is precisely on account of the latter fact that the treeshrews serve so well as models of what we can assume to closely represent a basic primate type, from which all other forms developed in diverging directions. At any rate, there is excellent justification for treating the modern treeshrews as 'living fossils'. In regard to this as well as all the other classes of primates to be introduced here, we must bypass most of the technical arguments of taxonomists and the detailed reports of palaeontologists in favour of a brief character-isation of typical recent species which can claim the greatest interest in a book on the life of primates.

Treeshrews are invariably small and agile animals, rarely living high above the ground, but preferably in the low forest growth of tropical south-east Asia, including the large islands of Sumatra, Java and Borneo. They have a slender build, long bushy tails and comparatively short limbs, especially the anterior ones (figure 5). The small hands and long feet each have five clawed digits, of which the first or medial ones are the shortest and not well suited for grasping. The head and skull of treeshrews also show most of the typical unspecialised characteristics of primitive mammals. The long facepart projects straight forward with the nostrils well in front of the receding chin, the mouth is strikingly large, the eyes and orbits are still laterally situated, and the occipital condyles lie at the extreme end of the skull, so that the head is carried fully in front of

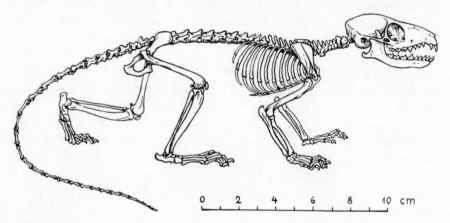

Figure 5   Skeleton of an adult treeshrew (*Tupaia glis*), showing the typical posture with flexed limbs.

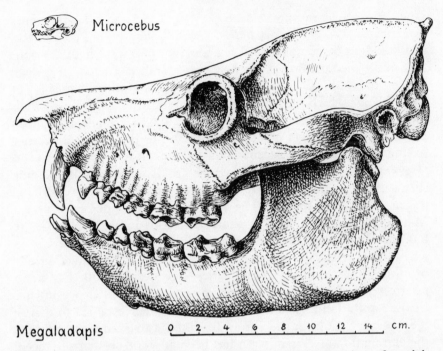

Microcebus

Megaladapis

0 2 4 6 8 10 12 14 cm.

Figure 6   Skulls of an adult *Megaladapis edwardsi* and for comparison of an adult *Microcebus myoxinus*, both drawn to the same scale, shown below.

the vertebral column. Of the primitive dental formula of three incisors, one canine, four premolars and three molars on each side of both jaws, only one incisor and one premolar have become eliminated in the recent Tupaiiformes, when all these teeth still existed in some of their early predecessors.

A second infraorder of Prosimiae, the Lemuriformes, has for a long time been restricted to Madagascar and the neighbouring Comoro islands, where it has evolved into remarkably many more or less different forms, which, in turn, can be subdivided further into several distinct families, subfamilies and genera as well as numerous species. Only few of the latter need be dealt with separately here, all the recent genera being listed in a table in the appendix. Almost all of these lemur-like types are thoroughly arboreal and many are nocturnal, their diet consisting of leaves, fruit and other vegetable substances, often extensively supplemented by insects. Most of the nocturnal species build nests of leaves in which they raise their often quite immaturely born offspring. In body size the Lemuriformes

17

range from the tiny mouse lemurs (*Microcebus*) and dwarf lemurs (*Cheirogaleus*) to the indri (*Indri*), weighing about six kilogrammes and standing on its long hind legs nearly one metre high. Various now extinct lemuriform types were considerably or even very much larger, most of all the several specialised subfossil species of *Megaladapis* which are regarded as representing a family of their own (figure 6). With the exception of the latter, all lemuriform prosimians have longer posterior than anterior limbs, this difference having become most accentuated in the closely related sifakas (*Propithecus*) and indris (*Indri*), which are distinguished by holding themselves upright in sitting, hopping bipedally on the ground and leaping from tree to tree. The indri is the only species of recent Malagasy lemurs to have a short tail, containing from eight to fourteen small vertebrae, whereas all the other species have anywhere from twenty to thirty-two tail vertebrae. The reduction of a primitive long face and snout varies considerably, but has never progressed as far among lemurs as in some of the other prosimians. In the family Lemuridae the dental formula corresponds to that of the treeshrews, but in the family Indriidae a marked trend towards the elimination of certain teeth has been at work, so that the recent species have only two premolars left on each side of both jaws and even the lower canines have become lost, though they still appear in the milk dentition.

An extreme reduction and unique specialisation of the dentition characterises the aye-aye, that aberrant Madagascar species, representing among the Lemuriformes a separate family of Daubentoniidae (figure 7). This animal lacks permanent lateral incisors, canines and premolars, except for a single small one of the latter on each side of the upper jaw. It is highly significant that in the milk dentition lateral incisors and upper canines still appear, but are not replaced by permanent ones. Due to the peculiarity of its huge incisors the aye-aye had been regarded as a rodent when first described by earlier naturalists. Its inclusion among the Lemuriformes, however, has become fully justified by the more recent discovery of its numerous basic primate characteristics. Besides being able to gnaw wood and open hard-shelled fruit with its sharp incisors, the aye-aye has developed extremely long and slender middle fingers, specialised for hooking the larvae of beetles out of holes in the bark. This conspicuous feature has earned it the German name of *Fingertier*. Aye-ayes are strictly nocturnal and exclusively

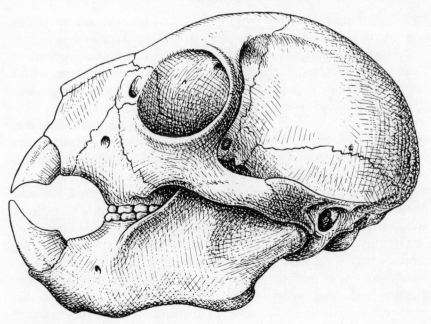

Figure 7   Skull of an adult female aye-aye.

arboreal animals, no larger than a small cat, and not in the least gregarious. A similar species, closely related to the only surviving one, may have become extinct quite recently and was considerably larger.

The third prosimian infraorder is known as Lorisiformes and contains two distinct families of Lorisidae and Galagidae. The first of these families has a wide distribution in tropical Africa and southern Asia, while the second one is limited to the African continent. All the Lorisidae are very slow and deliberate in most of their movements in dense forest, where they are active only at night. Their tails have become much reduced in length and number of vertebrae, and the second digits of the hands and feet also tend to be unusually short, while the first digits are long and well suited for remarkably powerful grasping. The large eyes of the Lorisidae are situated fairly far forward and close together in connection with the comparative shortening of the face. The front and hind limbs are of nearly equal length, and in the so-called slender loris they are so exceedingly thin for their length that it has been described as resembling a banana on stilts. The slender loris (*Loris*) and the slow loris (*Nycticebus*) form the two living Asiatic genera, and the pottos (*Perodicticus*) and angwantibos (*Arctocebus*) the African genera.

The African family of Galagidae, called bushbabies on account of their baby-like cries, contains a number of different species which some authors like to group into several separate genera. The largest species (*Galago crassicaudatus*) is hardly as big as a cat and the smallest, or dwarf galago ('*Hemigalago*' *demidovii*), not even half that size. These bushbabies are mainly, but not strictly, nocturnal, are very agile and equipped with long tails. Their dental formula corresponds to that of the Lorisidae, which is also the same as that of the living treeshrews. The Galagidae are distinguished by their phenomenal leaping ability, due to their long and strong hind limbs and, particularly to the great elongation of certain tarsal bones.

The tiny tarsiers from the Malay archipelago are the only living representatives of what is probably best regarded as the fourth prosimian infraorder, the Tarsiiformes, though the precise class-ification and designation of these highly specialised primates has been and still is being debated with unusual intensity by comparative anatomists. The modern tarsiers differ from all other prosimians in having lost the typical rhinarium and by having gained a partial bony partition between the orbit and temporal fossa as well as a new position of the foramen magnum and the occipital condyles which have shifted forward from the extreme rear end of the skull (figure 2). In these and various other quite significant respects the tarsiers have come to assume certain simian characteristics, but a great many add-itional features clearly group them with the prosimians. Their out-standing specialisations consist of the unique enlargement of their eyes and orbits as perfect adaptations to their nocturnal life, and in the extreme lengthening of their hind limbs, including the heel region of their long feet, in connection with their great leaping ability. The very long and partly naked tail serves as a prop in clinging upright to branches and in changing direction while jumping. Except for the enormous eyes, the face, jaws and particularly, the nasal chamber are much reduced, but the outer ears are very large and mobile, showing that these animals rely far more on hearing than smelling. Tarsiers greedily devour all sorts of insects as well as small reptiles, birds and even young mammals, which they catch with their hands and lift to their mouths, as do so many other prosimians. While clinging to a branch they can swivel the head 180° right or left by twisting movements in the inter-vertebral joints of the neck. This uncanny peculiarity has added to the superstitious dread which these harmless creatures inspire in most Malay natives.

Fossil remains of many prosimians have been found in deposits of early Tertiary age in North America, Europe and China. The great majority of them belong to the lemuriform group and many others to the tarsiiform one, as far as can be determined from the fragmentary evidence derived for the most part from teeth and skulls. It is highly interesting that these very early prosimians had such a wide distribution in northern territories, and had already diverged very markedly in many of their characteristics. It seems perfectly possible, therefore, and indeed quite likely, that the simian primates of the Old World and those of America had originated from different ancient stocks in these widely separated regions, but which stocks these were is as yet quite uncertain.

The second and younger suborder of primates – the Simiae – is in most respects more highly developed than the prosimians, but it has not become as much diversified as the latter, being in general a more compact and uniform group as far as its main anatomical characters are concerned. Of these only a few can be mentioned here as being typically simian. The brain has become not only larger in relation to body size, but has also been improved in various important ways, while the entire apparatus for the sense of smell has undergone further reduction. The forward shift of the eyes and with it binocular vision has been completed and the orbits are invariably separated from the temporal fossae by extensive bony partitions. The foramen magnum and the joint connecting the skull with the vertebral column, lie far forward in early stages of development, but subsequently shift toward the rear in varying degrees. The permanent dentition normally contains two incisors, one typical canine, two or three premolars, and two or three molars on each side of both jaws, which tend to migrate underneath the braincase. Grasping ability in the first digits is well developed in most species. Secondary sex differences exist generally in more ways and in more marked degrees than among prosimians. The duration of pre- and postnatal growth tends to become prolonged and the age of sexual maturity is reached later in all, but in especially the higher, simian primates.

The suborder Simiae can be subdivided into two readily distinguishable main groups, corresponding to their geographical distribution before the coming of man. All the monkeys of the New World are known under the technical name of platyrrhines and all monkeys of the Old World, apes and men as catarrhines, implying that they are differentiated by the possession of broad and narrow noses respectively. To be more precise, this distinction refers really to the nasal septum, since in the former group the nostrils are far apart and directed sideways, whereas in the latter they are usually close together and pointing forwards or downwards, as is indicated by the examples in figure 8. In their dentition platyrrhines have retained three premolars, whereas all catarrhines have only two (figure 9). Another constant distinction, readily seen in skulls of adults, exists in the bony tube, leading from the tympanic membrane to the outer opening of the ear in all catarrhines, in which it develops during postnatal growth, whereas in platyrrhines no real tube is formed at any age and the ear drum remains plainly visible. The tail plays an essential role in the life of all platyrrhines, it contains anywhere between twenty-one and thirty-four vertebrae, except in one form (*Cacajao*) in which it has been considerably reduced in length and number or segments (average thirteen). Among catarrhines, on the other hand, the tail varies enormously in its development, being very long in some species, but reduced to a mere stump or having even

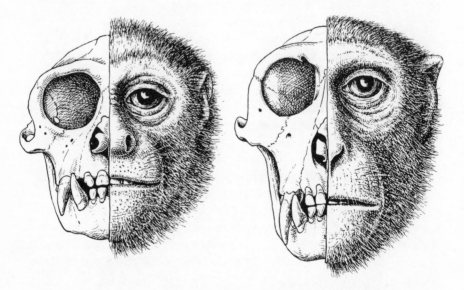

Figure 8   Sketches of the heads and skulls of an adult male platyrrhine *Cebus*, (left), and of an adult male catarrhine *Macaca*, (right), showing typical differences in the outer nose and the wing cartilages.

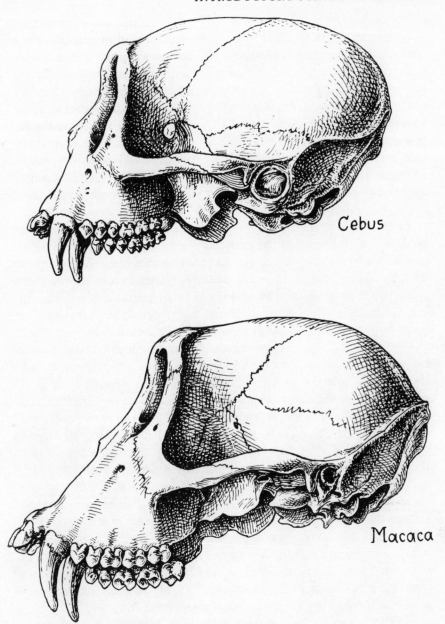

Figure 9   Side views (from slightly below) of typical skulls of adult male New and Old World monkeys (*Cebus capucinus* (top), and *Macaca sinica* (bottom)), showing the differences in the numbers of premolars, the relations between the bones in the temporal region and the formation of the ear opening.

23

disappeared completely in many others. It is only among the New World monkeys that there still occurs a single nocturnal genus (*Aotus*) with typically enlarged eyes and only among these Simiae is found a subfamily (*Callithricidae*) in which multiple offspring represents the rule. For these and many other detailed reasons platyrrhines can be regarded as being in general somewhat less advanced than the monkeys of the Old World. It is very remarkable, however, that there exist so many close similarities between platyrrhine and catarrhine monkeys in spite of their independent evolution during what was undoubtedly a long period of time. This fact can be explained only as due to parallel adaptations to very similar modes of life.

The American monkeys which are all placed in the one superfamily of Ceboidea can be separated into quite distinct families, called Callithricidae and Cebidae. The former, whose name 'callithrix' implies that they are 'beautiful-haired', include the many species of marmosets and tamarins which appear to differ strikingly, but are really quite uniformly built, except for their often very gay hairy coats and some details of their dentition, etc. That the Callithricidae are to be regarded as the most primitive and archaic types of platyrrhines with an entirely separate evolutionary history appears likely, but is still unverified, especially because no fossils have as yet been found for reliable help in solving this problem. The outstanding distinctions of this family are the possession of sharp claws on all digits, except the first toe, and the loss of third molars in all but one species (*Callimico goeldii*) which has recently been transferred from the Cebidae. The elimination of the last molar teeth is certainly a localised and minor specialisation which has become nearly universal in this family, while among the Cebidae a corresponding trend to reduce the last elements in the dental row is merely indicated by its sporadic manifestation in some species of *Cebus*, *Ateles*, etc. The well-functioning claws of the marmosets, which have given them the German name 'Krallenäffchen', may have developed as modifications of original nails in adaptation to the squirrel-like mode of arboreal locomotion of these very small primates, which have no effective grasping power in their first digits, but with their claws can secure firm holds on the bark of trees.

All Callithricidae have slender and flexible trunks with long lumbar regions and limbs of moderate length, the posterior ones being little in excess of the anterior ones in accordance with their

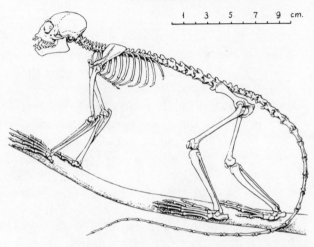

Figure 10   Skeleton of an adult tamarin showing the exact proportions in all parts.

typically quadrupedal locomotion. The tail is always well developed, the primitive brain and the eyes are proportionately large for reasons of allometry, and the occiput projects beyond the cranio-vertebral joint. The jaws are comparatively small and only moderately protruding. The shoulder blades lie low on the sides of the narrow chest and the slender hip bones connect with only one or two of the small sacral vertebrae (figure 10). All these conditions in general represent a good model of an unspecialised type of monkey skeleton.

The family of Cebidae contains at least ten different living genera, representing a remarkable variety of specialisations and of evolutionary progress, even though all these monkeys are entirely arboreal, live in a tropical or, at least, semi-tropical environment and have a chiefly vegetarian diet, merely supplemented by insects and occasionally other animals. All adult Cebidae surpass the marmosets in size, some of the largest male howler monkeys or female spider monkeys attaining body weights of up to ten kilogrammes. The closely related genera of *Aotus* and *Callicebus* are generally rated as the least advanced types, judging by certain conditions of their brains and other anatomical features, as well as by their behaviour. Also rarely seen in captivity are the many and often very decorative species of the genera *Pithecia*, *Chiropotes* and *Cacajao* which are medium-sized Cebidae with various localised specialisations, of which some will be mentioned later on. Most frequently seen in zoos and also today in research institutes, are the lively and attractive squirrel monkeys (*Saimiri*) and capuchin monkeys (*Cebus*), the latter having become specially well known for their often astonishing mental ability. The howler monkeys (*Alouatta*) are highly specialised in their vocal apparatus, particularly with regard to the enormous enlargement of the hyoid bone. In addition they have acquired a prehensile tail with a hairless tactile surface on the ventral side of the distal end. The latter extreme specialisation exists also in the remaining genera, the spider monkeys (*Ateles*) and woolly monkeys (*Lagothrix*), whose long tails serve almost like a 'fifth hand' in holding and even suspending the body as well as in reaching for objects. The last two types of monkeys belong to the largest platyrrhines and are probably the most intelligent ones. They are long-limbed experts in brachiating, but can also walk bipedally with apparent ease as well as quadrupedally, thereby resembling the gibbons of the Old World. Spider monkeys are furthermore distinguished by the loss of the phalangeal part of the thumb.

The simian primates of the Old World can be divided into two major groups, distinct since ancient times, namely the superfamilies of Cercopithecoidea and of Hominoidea. The first of these includes all of the few fossil and many different surviving catarrhine monkeys and the second the different man-like apes and man, together with their nearest relations of the past. The Cercopithecoidea are given superfamily rank to emphasise their roughly equivalent status to the distinct group of the highest primates, but for the former only one single family (Cercopithecidae) suffices, whereas the latter represent anywhere from two to five families according to the still differing views of modern taxonomists. Two subfamilies of Old World monkeys are usually distinguished, even though the entire family of Cercopithecidae is really quite uniform in nearly all of its main features. The many and mostly Asiatic species of macaques (*Macaca*), including the rare macaque- and baboon-like 'black ape' from Celebes, and the African baboons (*Papio*), geladas (*Theropithecus*), mangabeys (*Cercocebus*), guenons (*Cercopithecus*) and patas monkeys (*Erythrocebus*) are the well-known living forms of the subfamily Cercopithecinae. All the various Asiatic langurs (*Presbytis, Pygathrix*), the also Asiatic snub-nosed and long-nosed or proboscis monkeys (*Rhinopithecus, Simias, Nasalis*), which probably do not deserve to be kept in separate genera, and the African guerezas (*Colobus*) form the second subfamily Colobinae.

Besides the generally valid distinctions of catarrhines, of which some have already been mentioned, all Cercopithecidae have in common conspicuous ischial callosities which are thick horny plates fitting tightly over the bony tuberosities of the hip bones, on which the animals sit. These callosities develop long before birth in contrast to those found among the man-like apes, in which they appear later and remain thinner and smaller. The trunks of Old World monkeys are comparatively slender with narrow shoulders and hips and the chest is more deep than wide. The limbs are of nearly equal and moderate relative length and the first digits are short, rotated and fairly well suited for grasping, except for the thumbs of guerezas, which have become reduced to useless vestiges. The tail forms a long balancing rod in all species of guenons, mangabeys, langurs and guerezas, but in a great variety of macaques, baboons and the snub-nosed monkey *Simias* it has become very markedly shortened while in still other macaques and baboons, particularly the 'Celebes' and 'Barbary' apes, the tails have been reduced to mere stumps or knobs. All these monkeys are expert climbers, moving chiefly in quadrupedal fashion, but some macaques, baboons and the patas monkeys spend much more time on the ground than in trees, the last-named ones having become the fastest running primates. In many species the sexes show far greater secondary differences in size, dentition, hair, etc. than they do among the platyrrhine monkeys. This is specially pronounced in regard to the canine teeth of most Cercopithecinae which attain a very impressive size in males while remaining small in females.

The subfamily of Cercopithecinae is characterised not only by certain details of its dentition, but also by the possession of large cheek pouches as bilateral additions to the oral cavity in which food can be temporarily stored (figure 11). One of the chief anatomical distinctions of all Colobinae consists of their subdivided large stomachs which enable them to subsist almost exclusively on a diet of leaves and fruit. Even though the Colobinae are called 'Schlank-affen' in German, they are really not more slender than many Cercopithecinae; indeed, such forms as *Rhinopithecus* and *Nasalis* are just as robust as are some baboons. The body weight of fully grown individuals of the recent species of Cercopithecidae ranges all the way from less than two kilogrammes in the dwarf guenon (*Cercopithecus talapoin*) to nearly forty kilogrammes in male mandrills (*Papio sphinx*). Among fossil remains of Old World monkeys

27

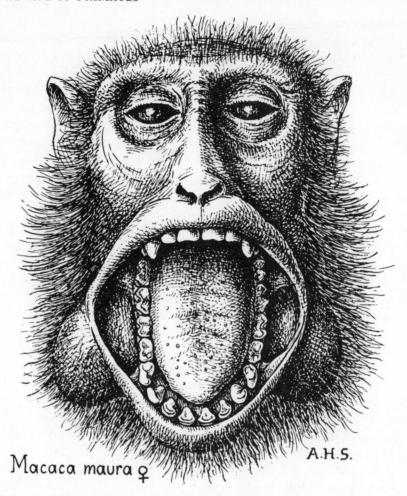

Macaca maura ♀   A.H.S.

Figure 11   Head of an adult macaque with mouth forced wide open and corners
pulled back to show cheek pouches which have been moderately extended.

there are some species which must have been still larger. Numerous
macaque- and baboon-like fossils from late Pliocene and Pleistocene
deposits have been found not only in Asia and Africa, but also in
Europe. These and even the few earlier forms of Cercopithecidae
already show the well defined characteristics of this group.

The superfamily of Hominoidea is of course of paramount
interest since it contains man. A great number and variety of fossils
have been discovered, which undoubtedly belong in this particular
group of primates, but no universal agreement has as yet been

28

reached regarding the interrelations and detailed classification of some of these finds, which will be briefly discussed later on. The recent Hominoidea are for sound reasons usually grouped in three families, namely the Hylobatidae for gibbons and siamangs, the Pongidae for the great apes, orang-utans, gorillas and chimpanzees, and the Hominidae for all the races of mankind. In general size these closely allied primates show a remarkable variety since most gibbons weigh less than six kilogrammes when fully grown, whereas some male gorillas can reach a weight of at least 175 kilogrammes. All these recent apes and man have many features in common, which separate them from the lower catarrhines. The trunk of the adults is stout with comparatively broad shoulders and hips and with the chest wider in its transverse than its antero-posterior diameter. The shoulderblades have shifted to the back and the vertebral column toward the breastbone, which itself is not only much wider than in monkeys, but also tends to become solidified through fusion of most of its intercostal segments. The combined number of thoracic and lumbar vertebrae is reduced from an original general average of nineteen to about eighteen in gibbons, seventeen in siamangs and man and even less in the great apes, especially in orang-utans. The number of sacral vertebrae, on the other hand, has increased from generally only three in monkeys to averages of anywhere from just below to well above five in all recent Hominoidea, which are furthermore distinguished by an extreme reduction in caudal segments and the complete loss of an outer tail. The brain has gained significantly in relative size, in the convolutions of its surface and in manifold internal perfections, which are well advanced in the pongids and have most extremely progressed in man. The limbs have become clearly longer in relation to the trunk, as a general hominoid specialisation which is most pronounced in the arms of gibbons and orang-utans and, less extremely, in the legs of man. Gestation and the postnatal period of growth have been prolonged and the ages of sexual maturity postponed, so that the intervals between succeeding generations surpass in hominoids those of all other primates. It may be mentioned also that the man-like apes seem to be unable to swim and that the great apes build nests in trees or on the ground.

The gibbons (*Hylobates* = literally 'tree-walker') are the most numerous of the man-like apes and their at least five different species still range over wide forested areas of south-east Asia. In general they have remained somewhat more primitive and monkey-like

29

than the other apes, their small size, moderately advanced brains and regular, thick ischial callosities exemplifying such characteristics. In regard to many other features, however, gibbons have acquired unique specialisations, such as an extreme density of hair and a great variability in the colour of their coats. Together with their larger cousins, the siamangs, gibbons are also distinguished by their extremely long arms, including the slender hands, and by possessing very long canine teeth in both sexes, which are of the same average body size.

Whether the rare siamangs deserve to be assigned to a separate genus (*Symphalangus*) is still debatable, but the fact that they approach the great apes in many ways much more closely than do any species of true gibbons should be recognised taxonomically. The single recent species of siamang lives in Sumatra and a slightly smaller race thereof has become isolated in a limited region of the Malay peninsula, but in the Pleistocene age a larger form of siamang had existed also at least in Java. The relative length of the limbs is not quite as extreme in siamangs as in gibbons and the density of their hair is not nearly as great. All Hylobatidae are remarkably versatile in their locomotion, being able to run upright on their feet alone, to swing underneath branches by their arms, or to climb quadrupedally with equal ease according to the nature of the support.

The orang-utan (*Pongo*) is the only living Asiatic member of the family Pongidae. Having formerly had a wide distribution on the mainland, it has become restricted to Borneo and a small part of Sumatra, where it leads an almost exclusively arboreal life, to which its limbs have become well adapted. Male orang-utans can attain weights of at least seventy-five kilogrammes, but females remain much smaller. In various other ways they have also acquired fully as marked secondary sex differences as have gorillas, in striking contrast to all other hominoids. Orang-utans are the only primates whose first toes have much degenerated and whose outer ears are extremely small. They have very long, but sparse hair of copper-red to yellow colour, huge throat pouches and most old males develop widely flaring cheek pads (figure 12). In contrast to nearly all other primates, orang-utans tend to be quite solitary and one rarely finds more than mother and child or two sexual partners in close association.

The chimpanzee (*Pan*) and the gorilla (*Gorilla*) represent the family of Pongidae in Africa, where they share the central tropical region between the west coast and the eastern chain of lakes. Both

Figure 12   Head of an adult male Sumatran orang-utan with light-yellow beard and lips posed for calling (from photograph of a living animal).

these apes are at home in dense forest, but chimpanzees are found also in some fairly open districts with just enough trees for safe retreats during the night. Chimpanzees and gorillas can climb readily and even swing by their arms alone, but generally spend much more time on the ground, particularly the heavy gorillas. In quadrupedal walking or running they support themselves on the middle segments of their flexed fingers. They are also well able to stand, walk and run bipedally, thereby having their hands freed to carry food and other objects, or to hold their newborn.

Chimpanzees are the smallest of the 'great apes', females averaging about forty kilogrammes in weight and males forty-five kilogrammes when fully grown, but their individual size varies extensively. The so-called 'dwarf chimpanzee' from the left side of the Congo river should better be referred to by its other name 'Bonobo' (*Pan paniscus*), because it is not smaller than many individuals among the several races of other chimpanzees (*Pan troglodytes*). Female chimpanzees are distinguished by their very conspicuous cyclic sex swelling of the pudendal region, but otherwise sex differences are not nearly as marked as they are in gorillas and orang-utans. Chimpanzees have exceptionally large outer ears, usually black hair of very scant density and skin colour ranging from nearly white to brown and black, often undergoing marked changes with advancing age. It is also noteworthy that they are more gregarious as well as more noisy than the other apes and that they have repeatedly been observed to catch and eat small mammals.

Gorillas are the largest of all living primates, particularly the males which become about twice as heavy as the females. The eastern subspecies (*Gorilla gorilla beringei*), often but inappropriately called 'mountain gorilla', differs somewhat in various respects from the western one. Some recent authors have placed the gorilla in one and the same genus *Pan* with the chimpanzee on account of the close similarity of their chromosomes and certain qualities of their blood, but this proposed 'lumping' does not take into account their great many marked dissimilarities in regard to other significant characteristics. Gorillas have developed such huge dentitions and jaws that their powerful muscles of mastication require for their attachment extraordinarily strong zygomatic arches and bony midsagittal crests. The latter, in turn, become covered by a thick pad of tough connective tissue which gives the head a peaked crown (figure 13). To balance the heavy head nearly in front of the spinal column has

Figure 13   Head of an adult male western gorilla (from photograph of a living animal).

called for an enormous nuchal musculature which is connected with a strong transverse bony crest on the occiput and with exceptionally long spinous processes of the cervical vertebrae. The trunk of the gorilla is far more voluminous and plump, particularly in breadth, than that of man and in old males the chest is nearly bare. These and

33

many further features give the adult gorilla its ferocious appearance and reputation, but actually it is a remarkably peaceful creature, using its incredible strength merely in self-defence. Corresponding to their size, gorillas have the largest brains of all nonhuman primates, but they do not seem to surpass the much smaller chimpanzees in intelligence.

The hominoid family of the Hominidae is today represented by man alone, who is the only catarrhine primate whose distribution extended to the New World for probably the past 15,000 years. With the early acquisition of an erect posture and bipedal locomotion and the much later enormous enlargement and perfection of the brain man gradually evolved his decisive culture and language and became able to adapt himself to all climates and to remarkably varied diets. Man is furthermore distinguished by the most extreme reduction of hair and the greatest prolongation of the postnatal periods of life. Regarding some of his bodily features, however, man has not become as highly specialised as have various other primates, as will be discussed in later chapters. With his far-reaching control of nature and ceaseless inventiveness, primed by typical simian curiosity, recent man has acquired the unique potentiality for overpopulating the earth.

Chapter 3

Origin and Distribution of Primates in Time and Space

THE EARLIEST fossil representatives of the order of primates have been discovered in the middle of the Palaeocene period and are therefore about seventy million years old according to modern estimates. This ancestral primate stock had become segregated from other archaic placental mammals by at least some of the few characters identifiable in their fragmentary remains. Several different types had already appeared among these primitive Palaeocene primates, which had all been found in North America and whose systematic position had not been undisputed since they still show some close resemblances to insectivores. Mainly for the latter reason it is generally agreed today that the very first primates must have risen from among early insectivores and most likely the types from which the later tree-shrews had developed, which some mammalogists still regard as insectivores, rather than primates.

During the late Palaeocene and the subsequent geological period, the Eocene, there rapidly appeared a surprisingly large variety of still very archaic prosimians in Europe and North America as well as finally also in southern Asia, but so far none have been found in Africa nor in South America. They were all small animals, most likely arboreal and with generally quite primitive conditions of skeletons and teeth. For instance, in the only two skulls of Palaeocene primates recovered in an adequately complete state, the orbital and temporal fossae are as yet undivided by a bony postorbital bar, which becomes a constant feature of all later prosimians. Many of these diversified forms seem to have died out towards the end of the Eocene, since only few can be traced to later evolutionary representatives. It is of great interest to find that already among the later Eocene primates types have appeared with larger brains, relative to their body bulk, than in any other contemporary mammals and that in some the visual sense must have predominated over the olfactory sense and that furthermore the face had become markedly shortened

35

and the eyes moved from a lateral to a more frontal direction. With these conditions some of the Eocene prosimians already foreshadow the same later and much more pronounced trends among higher primates. It is particularly the very widely distributed Eocene family of Omomyidae with its numerous genera which is most likely to have been the ancestral stock of the simian forms which must have started their evolutionary careers in the next geological epoch, the Oligocene. Fossil remains of primates from the latter period are few and far between and seemed to lack undoubted prosimian forms and to have been limited to North Africa before the very recent and fortunate discovery in Texas by J. A. Wilson of a well-preserved skull from the early Oligocene, which is regarded as a new prosimian and even a new omomyid genus, named *Rooneyia*. In general, however, the Oligocene, which lasted for about ten million years, still forms a sad gap in our knowledge of the history of primates. This, of course, does not imply that there occurred a period during which primates had been very scarce or had even become nearly extinct, but merely shows the role of chance in preserving remains as fossils and in finding them, which can differ widely in the various epochs as well as among the different orders of animals.

In the Miocene period, which succeeded the Oligocene and lasted probably fully twice as long, or some twenty-five million years, the suborder of prosimian primates has completely disappeared in Europe and North America, while the suborder Simiae is rather poorly represented by a few monkeys of the New World and even fewer of the Old World, but by a remarkable variety of Hominoidea, widely distributed over Africa, Europe and Asia. Even though the latter most progressive primates have formerly been excessively split into too many genera, particularly the so-called dryopithecines, it is quite evident that the Miocene hominoids did follow many diverging evolutionary trends and showed a pronounced radiation, similar to, but not nearly as marked as that of the Eocene prosimians.

From the succeeding geological period, the Pliocene with its estimated duration of about twelve million years, exceedingly few remains have been recovered so far of prosimians and of New World monkeys. On the other hand, a fair knowledge of the monkeys of the Old World really begins only with this epoch, in which this group is represented by several different genera. The Egyptian *Libypithecus* is possibly ancestral to the macaque and baboon group and *Meso-pithecus,* known through exceptionally large numbers of remains from Europe and the Near East, is most likely affiliated with the group of langurs, as also seems to be *Dolichopithecus* from a slightly later horizon of the Pliocene in France. Several further fossil finds of monkeys have been reported for this period in Europe and in India.

Hominoid primates of the Pliocene age must have been quite numerous and diversified in Europe and in parts of Asia according to the many reports on their remains. They are mostly close relations or direct descendants of Miocene forms, particularly of *Dryo-pithecus* and perhaps *Sivapithecus,* but include also such previously unknown types as *Oreopithecus* and *Ramapithecus* which are of special interest since they have been claimed to stand nearest to the evolutionary path leading to man. The former genus, known from unusually complete remains found in Italy, is now generally regarded as representing a separate hominoid family which became extinct in early Pliocene times. Based on recent re-evaluations of all the scant fragmentary remains of *Ramapithecus* from a wide geographic range, it seems quite reasonable to assign this hominoid a place among the early Hominidae on account of its short face, reduced front teeth and certain other detailed distinctions.

Most of the fossil primates from the youngest and shortest of the great geological periods, the Pleistocene, which lasted only about one to two million years, are forms differing comparatively little from corresponding genera of today. This is the case in tropical America as well as in the Old World with the exception of Madagascar, the isolated refuge of a multitude of lemuriform prosimians. In this huge island, nearly free of any larger predators until the late arrival of man, there have evolved at least six different subfamilies of which the Megaladapidae, Archaeolemurinae and many genera among the other larger groups have become completely extinct with, or even long after, the close of the Pleistocene. The astonishingly large differentiation in size and many other specialisations which had developed within this fauna resembles the great variety of marsupials which could evolve in the isolation of Australia. In their distribution most Pleistocene primates occupied much larger ranges than their descendants retained after this period. For instance, macaques still existed in many parts of Europe in interglacial times and Pleistocene orang-utans of often large size have been found in

Indochina, South China, and Java, besides Sumatra and Borneo, which constitute the only homes of their present relics. In the Pleistocene period China contained also another pongid of remarkable size, but without descendants, the so-called *Gigantopithecus*, of which many teeth and several mandibles have been recovered.

This brief outline of the history of primates according to the palaeontological record has been inserted here in such incomplete form because any fuller account would unavoidably call for a mass of technical details on teeth and bones, systematics and stratigraphy, as well as the frequently changing interpretations of many of the fragmentary finds. There have appeared numerous recent, excellent and full surveys or summaries of all the most important discoveries of fossil primates, particularly by such eminent experts as Remane (1956), Piveteau (1957), Thenius and Hofer (1960), Genet-Varcin (1963) and Simons (1963). In spite of all the devoted and painstaking work of these and a great many other scholars it has not yet become possible to give anything like a rounded off account of the history of primates as revealed by their remains from all geological periods and all parts of their changing distribution. Much must be left to speculation because primates have been chiefly arboreal animals, at home in mostly tropical forests, where the chances for fossilisation are far less favourable than they are in the case of terrestrial or aquatic animals especially of temperate zones. It is encouraging, however, that a far greater number and variety of fossil primates has come to light in recent years than in the earlier decades of fossil-hunting, so that the picture of primate evolution throughout the ages is quite rapidly gaining firmer support for many places and periods.

The *geographical distribution of the recent primates* would be much more puzzling than it is, were it not for the palaeontological evidence and our knowledge of the limitations in the spread of primates and in their adaptability. The accompanying partial map of the world (figure 14) shows the distribution of the recent nonhuman primates, except the treeshrews and great apes, which will be discussed below. Only tropical and parts of the temperate regions were required for this illustration. Today there are no more wild nonhuman primates in North America nor in northern Asia and none in Europe (except for the possibly reintroduced Gibraltar macaque), while there never have been any in Australia, New Guinea and all the islands of the Pacific Ocean. All prosimians have disappeared long ago in the

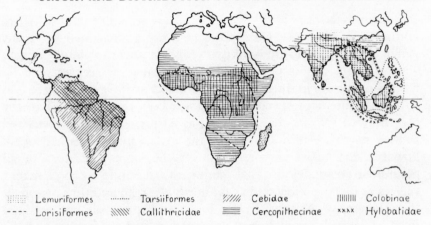

▦ Lemuriformes	⋯⋯ Tarsiiformes	▨ Cebidae	⦀ Colobinae
---- Lorisiformes	⧄ Callithricidae	☰ Cercopithecinae	××× Hylobatidae

Figure 14   Geographical distribution of the recent nonhuman primates.

entire New World, where the platyrrhine monkeys have been the only representatives since the early Pliocene period, limited to the more or less tropical forests between Mexico and Argentine. The small marmosets have a much more restricted distribution than some of the more adaptable Cebidae. The high chain of Andean mountains has formed an effective western boundary for all platyrrhines, except for a few Central American regions with only low hills. Since no ground-living platyrrhines exist today or have been found as fossils, the existence of forests is of the greatest significance for the distribution of all these varied monkeys of America.

In the Old World many of the major groups of recent nonhuman primates are represented in Africa as well as in southern Asia. The still very numerous genera of the Lemuriformes have become limited to Madagascar and the Tupaiiformes are found today on the mainland of Asia from India to the Malay peninsula and on all the large south Asiatic islands, except Celebes. The tarsiers survive only on the latter islands, including Celebes and the southern part of the Philippines, but not Java. Of the family Lorisidae the two genera *Loris* and *Nycticebus* live in southern Asia and the two other genera *Perodicticus* and *Arctocebus* are at home in Africa where also all Galagidae are found. It is seen, therefore, that the present descendants of some of the many forms of early prosimians, chiefly from North America and Europe, are making their last stand today in the tropical parts of Africa and Asia, and have fared best in their later isolation on such islands as Madagascar, Ceylon, Borneo, etc.

39

Both subfamilies of Old World monkeys are still found in Africa and in Asia, but the Cercopithecinae are represented by many genera in Africa and by only two (*Macaca* and the closely related 'Cynopithecus') in Asia, whereas the Colobinae are represented by numerous genera in the latter continent and by only one (*Colobus*) in Africa. Of the many species of macaques only the so-called Barbary ape (*Macaca sylvana*) has remained in Africa, where it is restricted to the north-western corner in the Atlas mountains. In Asia, however, macaques have spread over the largest territory of all nonhuman primates in that continent, reaching with different species as high as Tibet and as far north as Japan and thriving on all the other major Asiatic islands. The apparently wide gap between the African and Asiatic homes of macaques had not yet been nearly as marked in Pleistocene times as later on, as is indicated by fossil finds from Europe and Egypt. The largest of the catarrhine monkeys, the various baboons, are today exclusively African, ranging all the way from some mountainous areas in the Sahara to the southern-most tip of the continent. Pleistocene baboons, however, have been found in India as well as in Africa. At least some species of baboons seem to be exceptionally successful in surviving alongside the rapid advances of human civilisation in east and south Africa. As mostly terrestrial quadrupeds they are not bound to forests, their males are formidable fighters and their large social groups provide efficient watchfulness, so that they have become 'pests' in some agricultural districts. Some other African monkeys have also developed such remarkable independence of forests that they can inhabit regions which are decidedly unsuitable for their arboreal cousins. These are the geladas of the highlands of Ethiopia and parts of the Sudan and the swift patas monkeys of the open savanna country across Africa well north of the equator.

The varied ecological factors influencing the complicated distri-bution of the mangabeys and guenons have been carefully surveyed by Tappen (1960), who showed the preferences of the many species for different types of forest, corresponding to similar conditions prevailing among guerezas. While it used to be very puzzling that so many closely related species of both these groups of monkeys lived in the same districts and, indeed, had been collected in identical localities, it could recently be shown that every species is adapted to its own particular life-zone, such as deep primary forest, edges of big forests, or marginal forest along rivers, secondary growth and either the lower, poorly lighted parts of forests or preferably the denser zone of tree-tops. Certain species, finally, are found almost exclusively in swampy forest, such as the so-called Allen's swamp-monkey of the Congo basin or the proboscis monkey of the Bornean mangrove swamps.

The Asiatic Colobinae, represented by the many species of langurs and the few of their closely related genera, are all decidedly arboreal, though some live in deciduous forests and must come to the ground regularly during dry seasons or in the winters of northern India or northwestern China. As shown by figure 14, their distribution coincides largely with that of the Asiatic macaques, but is not as extensive. The remaining Colobinae, the guerezas, range from coast to coast right across tropical Africa below the Sahara, being strictly bound to forested regions from sea-level to high up on mountain slopes.

The habitats of the recent man-like apes are the tropical forests of the Old World. The small gibbons still flourish on the mainland from Burma to the south-western borders of China and from the tip of the Malay peninsula to the mountains of northern Thailand. They are also at home in Sumatra, Java and Borneo, but not in Celebes nor the Philippines. The somewhat larger siamangs are limited to parts of Malaya and to Sumatra, where they occur in the same forests as the local true gibbons. Borneo and a limited district of Sumatra are the last refuges of the orang-utans and even here they

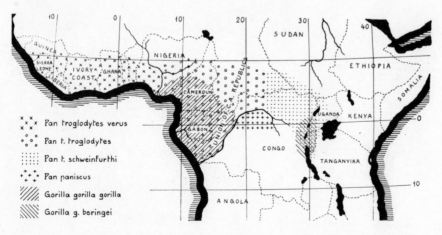

Figure 15   Geographical distribution of chimpanzees and gorillas in equatorial Africa.

have become very scarce in many extensive forests which would seem to be most suitable for them.

The recent geographic distribution of the African great apes is shown in figure 15, from which it is evident that the range of chimpanzees is considerably wider than that of gorillas, whose western and eastern races have apparently become isolated from each other in recent times. It is significant also that the two different species of chimpanzees are separated by the wide Congo river according to nearly all reports, while the three subspecies of *Pan troglodytes*, recognised by most authors, have a practically continuous general distribution and often are not readily distinguished on account of their great variability. Chimpanzees as well as gorillas range from sea-level to altitudes of at least 8,000 feet, where frost is not uncommon at night.

In striking contrast to all other primates, man has gained a world-wide distribution toward the close of the Pleistocene period after which he invaded the American and Australian continents. The regions of origin of the genus *Homo* are still unknown, but most certainly have been somewhere in Africa or Asia and here most likely outside of any densely forested area.

Chapter 4

Factors Determining Distribution

THE GEOGRAPHICAL distribution of primates is first of all dependent upon their adaptability to the local climate at all seasons and to other environmental conditions including any natural barriers, such as water, mountains, food supply and lack of means of protection from predators. While the large majority of nonhuman primates thrive best in more or less humid tropical and semi-tropical regions, there are many forms of catarrhines which seem to be easily able to withstand cold temperatures at night or even during wintry seasons. As already mentioned, chimpanzees and gorillas can live at elevations with frequent light frost and gibbons, also, range surprisingly high on such lofty mountains as Kinabalu in Borneo or Doi Angka in Thailand. Some species of macaques, baboons, geladas and the rare *Rhinopithecus* are found in regions where man requires clothing and fire to keep warm. Japanese macaques manage to survive even severe winters by eating the bark of certain trees, by seeking sheltered, sunny slopes and thus avoiding the deepest snow, and by sleeping in tight huddles of entire bands, according to a careful study by Wada (1964). In many modern zoos and research institutes it has been learned that numerous species of Old World primates do thrive out-of-doors even in cold winters, provided they have ample space for exercise, some adequate shelter is available for rest, and they have become accustomed gradually. That this ability to withstand cold is not dependent upon the development of the coat of hair is evident from the fact that there exists an extreme difference in hair density between gibbons and chimpanzees and that the hairy mantles of hamadryas baboons and geladas are limited to only the front of the body and are merely indicated in females of these species. Furthermore, orang-utans, gibbons and most American monkeys possess either very long or else very dense coats of hair, yet live in steaming equatorial jungles, where man prefers the scantiest of clothing. Subcutaneous fat cannot play an effective role in the protection against cold among wild primates, as it does in such other mammals, as whales, seals and walruses, because no noteworthy well-distributed amount of fat has ever been found by the writer in any of the great many monkeys and apes he has skinned, except in some overfed captive specimens which had had far too little exercise. It is not surprising, therefore, that moderately cold mountain ranges form no hindrance to the spread of many primates, provided there is no accompanying radical change in vegetation, particularly in forest and its suitable food.

Very high temperatures are apparently not well tolerated by at least many Old World primates. Though they will often ascend to the tree-tops for the warmth of the rising sun and for drying the nightly dew on their fur, they will also seek midday shade and actually suffer in bright equatorial sunlight. In temperatures approaching 100° Fahrenheit macaques have become unconscious and have even died later on, after only their palms and soles seem to have become moist with perspiration. Captive gorillas, forced to remain in the midday sun, have been known to die from tropical heat stroke.

Since most primates are arboreal, their range is usually limited by the extent of forests. In countries with long dry seasons forest fires can destroy or divide parts of the home range of primates and thereby isolate local populations, at least for long periods. In many parts of Africa and Asia, where the rapidly increasing human population deliberately burns large tracts of forest to gain land for its own use, the natural home of nonhuman primates has been sadly reduced in recent years. Even though a large share of the food of many monkeys is obtained on the forest floor, rather than up in the trees, the latter do provide the vital protection against the larger predators, especially at night. With the single exception of the night monkey *Aotus,* all monkeys and apes are diurnal and comparatively helpless during the dark hours of the long tropical nights. Adequate nocturnal protection is therefore a very decisive factor for primate distribution. Most species are reasonably safe in trees, where they are hidden in foliage, have left no trail of scent and can withdraw to branches too thin for the weight of such arboreal predators as leopards, ocelots or the big snakes. Even such largely terrestrial species as macaques and baboons will promptly take to trees when danger threatens and for safe sleep. A variety of these monkeys has become well adapted to the protection of steep rocky cliffs on which they huddle together every night. On the rock of Gibraltar one can admire the astonishing mountaineering ability of the local macaques and the Swiss zoologists Kummer and Kurt (1963) have described and pictured the dizzying rock faces which the Ethiopian baboons have selected as nocturnal retreats. The powerful old male gorillas are the only nonhuman primates which scorn any such protection for the night by usually sleeping on their crude nests at the *base* of some big tree.

Water of moderate extent or depth seems to form no effective barrier for many primates. At least macaques, baboons, mangabeys, guenons, proboscis monkeys, capuchin monkeys and howler monkeys have all been observed to swim remarkably well at times even when there was no necessity for it. The crab-eating macaques have gained their name from the fact that those living at the sea-shore regularly forage at low tide, catching and eating small marine animals. One group of the intensively studied Japanese macaques has acquired the habit of carrying food, scattered for them on the sandy beach, into the salt water for washing, wading upright with the food held in their hands. The writer has repeatedly observed hamadryas baboons, geladas and rhesus monkeys swim with ease and apparent enjoyment in the deep water of the moats surrounding their enclosures in zoos. Proboscis monkeys, which live in the vast mangrove swamps of the Bornean coast, not only swim frequently and well in dog-fashion, but have also been seen to swim even under water for considerable distances and to dive readily from tall trees, hitting the water flat on their chests with resounding splashes (Kern, 1964). Of at least some species of guenons it has also become known that they take to water readily and can swim well, quite far and sometimes even under water. On the other hand, the man-like apes, according to all reports, seem to avoid getting into water and to be unable to swim when they do happen to fall in. For instance, an adult male gorilla in a New York zoo slipped into the water of the moat around the outside enclosure and drowned without any apparent struggle, a chimpanzee drowned quickly in the moat of the new ape-house of the Antwerp zoo, and a gibbon in the London zoo drowned even in very shallow water at the bottom of its large cage. On the other hand, it has also been reported that a few of the many chimpanzees at the zoo in Chester did finally get across the water of broad moats when highly excited.

The frequent occurrence of monkeys on islands far off-shore, such as Zanzibar, Fernando Po, Formosa, Hainan, Trinidad and Coiba, is probably explained by the geological history of these islands, which seem to be the remaining parts of former peninsulas. It is of great interest that most of these primates which had become isolated on islands differ in varying degrees from their nearest relations on the mainland by having developed some distinct racial or even specific characters.

Primates, naturally, do require water as part of their diet and during dry seasons they must remain near some source of water or migrate to it. In tropical rain-forests, however, they never have to leave the trees to obtain water, which collects in cupped leaves or knotholes and drips from every branch during the regular showers. In humid zones the heavy dew will also provide an ample supply of water without it having to be sought on the ground. All wild monkeys and apes must spend a very large part of their daily activity in obtaining nourishment, of which they rarely find adequate amounts without wandering or climbing from place to place. The distances they have to cover in their search for food vary with the type of environment and the season, but do not exceed the limits necessary for collecting what will satisfy their hunger. In their home range primates know amazingly well where and when the best sources of food can be found and hence they are normally most reluctant to venture beyond their accustomed territory, as long as it will continue to feed its occupants. While growing up, all primates soon learn particular food preferences from their elders and become bound to the environment supplying them with their specific menus. With advancing age monkeys and apes become generally much more hesitant in trying any new food, as every observant zoo-keeper can attest. Still, there seem to occur marked individual differences in this respect and, apparently, also local group differences. For instance, it was found that in some regions baboons have started to kill and eat small mammals and that some local chimpanzees have acquired the habit of eating termites as well as much larger animals. Certain marmosets have been reported as catching small fishes and some macaques as having become addicted to collect all sorts of marine food.

A very effective factor in the distribution of primates exists undoubtedly in the local prevalence of infectious diseases and of the vectors transmitting them, such as mosquitos, tsetse flies, etc. As will be discussed in a later chapter, nonhuman primates do suffer from malaria, yellow fever, filariasis, etc. and their populations can thereby become decimated after local flare-ups. Microscopic parasites of great variety are far more potent in checking the numbers of wild primates than all the larger reptilian, avian and mammalian predators together, except for man. Monkeys and apes are good to eat and comparatively easy to hunt with blow-pipe, crossbow, shotgun, snare or trap. The majority of human natives, sharing the ranges of their nonhuman relatives, do kill the latter gladly for their meat, or in some cases for their decorative skins, and partly also in protection of man's plantations. In recent years civilised man has been trapping live monkeys and apes in such incredible numbers for research purposes and zoos that some species are approaching extinction, having already disappeared from large parts of their former ranges. Many well-meant local and international regulations and laws for the protection of specially threatened species have been proposed, accepted and signed, but the effective enforcement of these is an almost hopeless task in many tropical countries. Attempts to control the trade in live nonhuman primates through official permits for their exportation and sale have not stopped the primitive, wasteful and cruel methods of natives in capturing and transporting animals which they can dispose of to smugglers for steadily rising prices. Chiefly young animals are being obtained in this way, their mothers having been killed, and many of the former also perish before they reach proper care. These sad facts appertain particularly to the man-like apes which are most in need of protection, since their comparatively slow growth and long interval between succeeding generations prevents rapid replacements in their decreasing populations.

The most potent factor of man's manifold interferences with the natural distribution and survival of other primates is his continual encroachment on the forest home of the latter. The extensive and steadily more effective destruction of forests with fire, power-saws and bulldozers to gain more grazing and arable land, to build wide highways and human settlements, as well as the inundation of vast tracts of land through huge dams, not only diminishes the remaining areas of forest, but breaks them up into separate lots, forming unnatural conditions for the best population maintenance of many wild species.

Chapter 5

Posture and Locomotion

IN DESCRIBING the different kinds of locomotion of primates it must first of all be pointed out that these chiefly arboreal animals have limbs designed for climbing in trees and have basically retained a corresponding construction. Nevertheless, all primates are also fully capable of moving on level ground and, indeed, many different species do lead a largely terrestrial life in daytime. Among prosimians *Lemur catta* is one such exception, inasmuch as these graceful animals do spend much time on the ground in search of food or in patrolling their territory. Platyrrhines leave their arboreal homes very rarely, capuchin monkeys being the only ones which in some places will come to the ground with considerable frequency. Among catarrhines terrestrial life has become much more common, though of all the Colobinae only the large *entellus* langurs spend much time on the forest floor. The Cercopithecinae, however, contain remarkably many species which have become well adapted to life in open districts with few or even no trees. Baboons, geladas and patas monkeys are outstanding examples of this and various macaques and the vervet guenons have become at least semi-terrestrial in some regions. Finally, among the Hominoidea, the gorillas move on the ground far more than in the trees, even though they are decidedly forest-dwellers. Even chimpanzees, when pursued, invariably come to the ground when fleeing. The extreme abandonment of arboreal life has been adopted long ago by man.

Primates, like the great majority of other mammals, are fundamentally quadrupedal by carrying their bodies on all four limbs, which are of nearly equal functional length in unspecialised forms, so that on level ground the trunk is suspended between the supporting extremities in a practically horizontal direction. These general conditions have been retained with little change in treeshrews as well as a variety of other prosimians and in many monkeys, such as marmosets, howler monkeys, macaques, baboons and guenons (figures 5 and 10). Many evolutionary modifications have been tried out in the limbs of primates for more or less specialised modes of climbing, leaping and other forms of locomotion. In some species the hind limbs have become markedly lengthened and strengthened to take over much more than half the moving force in climbing and, particularly, in propelling the body through space in jumping from one tree to another. This is most evident in sifakas, indris, bush-babies and tarsiers, which can leap for incredible distances, often equalling a great many times their own body length. In tarsiers, the champion jumpers of all primates, the total length of the lower limbs amounts to as much as 190 per cent of the length of the trunk, whereas in man on a general average to only 169 per cent. As will be discussed later on, the infinitely better jumping ability of tarsiers and some other nonhuman primates than that of man is due also to special adaptations in the feet. The subfossil giant prosimian

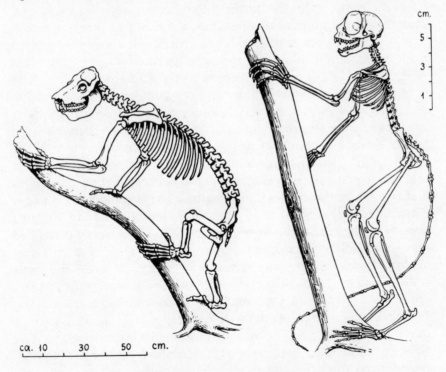

Figure 16 The proportionately shortest and longest hind limbs among primates in a *Megaladapis* (after reconstruction by Lorenz von Liburnau, 1905) and in a tarsier. The scale for the subfossil skeleton on the left is only approximate, the other scale is exact.

Megaladapis had its hind legs changed in the opposite sense, since they are so uniquely short that they could certainly not have served for jumping, but merely for very slow progression, probably much as in pottos. From the wealth of *Megaladapis* bones which have been recovered one can reliably reconstruct the entire skeleton and calculate that the percentage relation between the combined length of the femur and tibia and the approximate length of the trunk was only about 62, which is by far the lowest value of any primate and contrasts enormously with the highest corresponding value of 171 in tarsiers (Schultz, 1953 and 1954). These striking differences in the extremely specialised limb proportions of these prosimians are at once apparent in figure 16.

In a very considerable variety of simian primates the upper limbs are the ones which have become greatly lengthened and this not only in relation to the size of the lower limbs, but also to that of the trunk. All of these species have acquired the ability to progress swinging by their long arms underneath branches, a form of locomotion which has been termed brachiation. This has reached its extreme development in gibbons, in which the total arm length averages 243 per cent of the trunk length and 165 per cent of the total leg length. The corresponding values of siamangs are very similar with 233 and 178 respectively. In adult orang-utans the total arm length averages exactly twice the length of the trunk and 172 per cent of the total leg length. Among all monkeys the spider monkeys come nearest to the Asiatic apes in brachiating ability and in the relative length of their arms, though the latter averages only 180 per cent of the trunk length and 137 per cent of the total leg length. The elongation of the upper extremities of these best of all brachiators has affected the forearms even more than the upper arms, but this is not necessarily correlated with brachiation since many other primates also have forearms slightly longer than their upper arms. Though even the most pronounced of all brachiators, the gibbons and siamangs, can and do climb readily with all four limbs and walk on level ground and branches either quadrupedally or bipedally, they as well as most other brachiating species are distinguished by their astounding ability to throw their bodies through space by means of the momentum derived from the action of their arms alone and to catch hold again on some other support with only their hands. In the many monkeys which can brachiate to some degree and which can jump to perfection the hind legs usually provide the propelling force for the take-off in a jump, or at least contribute largely to it. The admirable grace with which gibbons can brachiate at high speed had apparently not become perfected until fairly recently, since no fossil finds show as yet such extraordinarily long and slender upper limb bones, as distinguish the recent species (figure 17). With brachiating locomotion the trunk, naturally, is held in an upright position, which is of great importance for many anatomical adaptations. The erectness of the trunk, however, is not at all typical of brachiating primates only, but is really maintained by all primates with widely varying degrees of completeness during some phases of arboreal locomotion, as well as in sitting which forms a considerable part of their daily life. When at rest tarsiers, indris, sifakas and still other springers usually cling to trees with their trunks and necks held vertically and their long hind legs sharply bent in the hips and knees.

The frequency with which brachiation is used depends on the environmental conditions for locomotion. On solid, thick and more or less level branches even gibbons, orang-utans and spider monkeys progress on top quad-

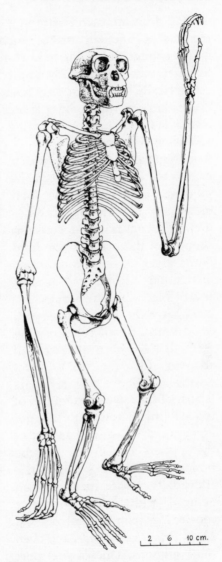

2   6   10 cm.

Figure 17   Skeleton of an adult male siamang. Note that in upright position the arms reach to the ground, even if the legs were straight.

rupedally or even bipedally. On the thinner, peripheral branches, too weak to carry extra weight without bending, balancing on top becomes very difficult so that the animal can safely proceed only by suspending itself underneath such elastic supports, swinging by its arms. None of the great apes actually makes much use of its brachiating ability in its native habitat, according to most recent reports. For instance, Dr G. Schaller, who has carefully studied all three pongids in the field, has categorically stated that only orang-utans brachiate occasionally and that he has never seen chimpanzees and gorillas to do so in the wild, he concluded that 'All great apes are essentially quadrupedal climbers.'

The upright posture is most closely and necessarily connected with bipedal locomotion, which has been a much more consequential specialisation than bimanual, brachiating locomotion. The natural and constant support of the body on the hind limbs alone is one of the main distinctions of man and as such must have become perfected comparatively early, inasmuch as all known fossil types of man seem to have been already well adapted for the fully erect posture. The proportionate length of the lower extremities of bipedal man, however, has not increased as much as has that of the upper extremities of the typical brachiating primates. Indeed, there are many human beings with proportionately shorter legs than have all tarsiers, *Loris, Propithecus, Indri* and even some gibbons. One of the most decisive advantages of bipedal locomotion is the liberation of the upper extremities for new functions during locomotion itself, especially the transportation of objects carried in the hands. The long-legged arboreal primates, like tarsiers, require

Figure 18 Female rhesus monkey standing fully upright (after a photograph in Frankfurt Zoo by Mr R. Lorenz).

52

Figure 19   Adult spider monkey and female gorilla walking upright on level ground, the former after a photograph (against the light) by Dr E. Erikson and the latter after a sketch by the late animal painter Charles R. Knight (original sketch published by Schultz, 1950).

their hands in climbing and in catching hold after a leap. Most monkeys and all apes are quite capable of assuming a fully erect posture, the former usually for merely a quick orientation from the elevated level of vision, gained by this posture (figure 18), and the latter frequently also for a limited bipedal walk or run (figure 19), occasionally dictated by a desire to have the hands free for fighting, for carrying some large food item or for supporting the feeble offspring (figure 20). Even though the upright posture and bipedal locomotion among nonhuman primates is not nearly as perfect as, nor strictly comparable in all details to, that of man, its widespread occurrence is most significant for our understanding of hominid evolution.

53

Figure 20   Female orang-utan
carrying its young in both of its
hands (after a photograph
of a captive animal).

Orang-utan

A.H.S.

There exist many transitions between these three basic types of
posture and mode of locomotion, which are not simply quadrupedal,
bimanual and bipedal ones, but are additionally differentiated by
various significant and detailed specialisations. Furthermore,
posture and locomotion change with individual development and
these changes can differ widely in the relative ages at which they
occur. Early infants of all primates, including man, pass through a
quadrupedal stage in their first attempts at locomotion, but can
support themselves bimanually remarkably well, due to the pre-
cocious development of the grasp-reflex of their hands. Bipedal
locomotion is clearly a secondary acquisition of later development,

54

perfected only by the end of infantile life in bipedal man, when many other primates also acquire the ability to stand and walk erect for shorter periods. Incidentally, it must here also be pointed out that the old and widely held notion of a *semi*-erect posture in standing on the feet alone is a mechanical impossibility and that any monkey, ape or man must stand fully erect to keep the centre of gravity of its body directly above the supporting feet, whereby the knees and hips may or may not be flexed and a tail may help in balancing. While

Figure 21 Quadrupedal postures in an adult female pigtailed macaque (after a photograph), a girl less than two years old (after Hrdlička, 1927), an adult female chimpanzee with marked sexual swelling (after Yerkes, 1943) and a male gorilla, 7½ months old (after Reichenow, 1921).

55

nonhuman primates can maintain such postures for a very limited time only, the human child gradually learns to stand and walk upright with increasing duration. Healthy chimpanzees learn to stand erect without any support at an average age of nine months, whereas human babies rarely before they are thirteen months old.

Quadrupedal locomotion does not always have to be literally four-footed, since many monkeys and apes do at times carry food or their helpless young in one of their hands while walking or even running with ease on three limbs only. Nearly all arboreal species place the entire hand flat on the support in walking, but the more or less terrestrial forms rest their weight on the extended fingers with the palm lifted off the ground, thus using the metacarpal portion as elongation of the forearm, whereby the head and visual field becomes correspondingly raised. The macaque in figure 21 is an example of the latter mode of supporting the body on the metacarpo-phalangeal joints of the digits II to V with the fingers hyperextended. The human child in the same figure shows the weight being carried on the carpus with the palm touching the ground, as in a great many nonhuman species, including the gibbons and siamangs. The chimpanzee and gorilla in figure 21 rest their weight on the knuckles, thus adding even the basal phalanges to the length of the supporting arms. The latter mode of usage of the hands in quadrupedal walking is typical of the African great apes and this from their earliest attempts at crawling throughout later life. It represents a specialisation by which the comparatively short flexor muscles of the fingers automatically bend the digits as soon as the middle part of the hand is dorsoflexed, so that the animals can also hang by their fingers alone without undue muscular fatigue. With this walking on the knuckles the skin over the back of the hairless middle segments of the fingers II to V has developed simple cutaneous ridges, comparable to the more elaborate dermatoglyphics on the palmar and plantar surfaces. In the orang-utan the mode of using the hands in walking resembles that of its African cousins in also usually having its fingers flexed. This, however, is more in the form of a clenched fist, rather awkwardly twisted sideways, and rarely like real knuckle-walking. The monkeys which bend back their fingers at right angles in walking seem to have acquired also much stronger legs than arms and, consequently, carry more weight on their feet than on their slender bent fingers. Probably for this reason the mothers of these species let their offspring preferably ride on their

Figure 22   Female hamadryas baboon carrying its young on its back (after a photograph of a captive animal).

backs over the pelvis rather than in front, as shown by the example of the hamadryas baboon in figure 22.

The terminal parts of the limbs of arboreal mammals, the digits, have the important function of providing a reliable hold on the supporting surface. In treeshrews and marmosets most or all of the nails are claw-shaped and function as sharp hooks in holding the light weight of these small primates on the bark of trees, as in so many other small mammals, birds and reptiles. In the great majority

57

of primates the digits bear more or less flat nails which are of no aid in holding the body in climbing to its support. The latter purpose is effectively accomplished by flexing the digits around a branch and also by grasping the branch between spread or even opposed digits. Such digital grasping has developed various specialised forms in different primates. Originally and still very frequently it is the first

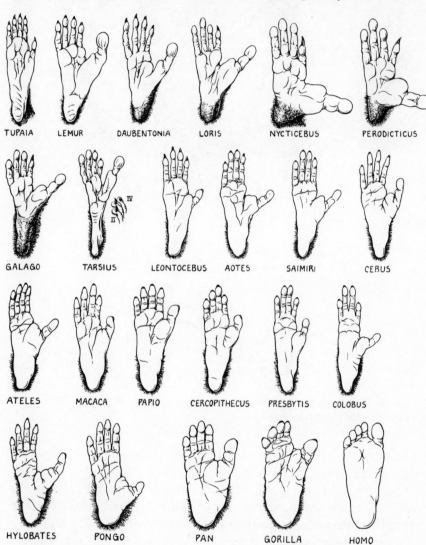

Figure 23   Right feet of some adult primates, all reduced to same length (after Schultz, 1956).

digit, especially of the feet, which is opposed to the lateral ones and this by mere spreading or by rotation and true opposition, so that the first digit really comes to face the others. Among prosimians the first toes as well as fingers act in grasping chiefly by spreading alone in such types as treeshrews and tarsiers, and by extreme abduction *and* rotation in some others, such as slow loris and pottos,

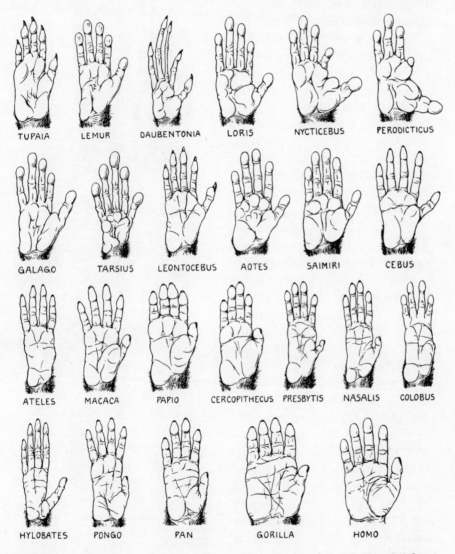

Figure 24 Right hands of some adult primates, all reduced to same length (after Schultz, 1956).

which have comparatively long and strong great toes and thumbs (figures 23 and 24). The thumb of New World monkeys is not nearly as well adapted to effective grasping as that of the Old World primates because it is at best very little rotated and branches from the palm near the base of the index finger (figure 24). These unspecialised conditions still exist among catarrhines in embryonic stages, but quickly change through the subsequent rotation of the thumb by about 90° and a simultaneous migration of the 'free' thumb in a proximal direction (plate 6). With these ontogenetic alterations the tips of the first and second (or other) fingers can be opposed like the jaws of pliers for a firm grip as well as for delicate manipulations. That the thumb is not always essential for arboreal locomotion is shown by the fact that it has become lost, except for small and useless remnants, hidden in the palm, in spider monkeys and quite independently also in guerezas. Both of these monkeys are excellently adapted to their swift arboreal movements, though they hold themselves by merely flexing the fingers II to V around branches. In many other monkeys, especially most langurs (figure 25), the thumb is pro-portionately too short and slender for strong gripping and the exceptionally long thumbs of gibbons are usually folded across the palm during rapid brachiating

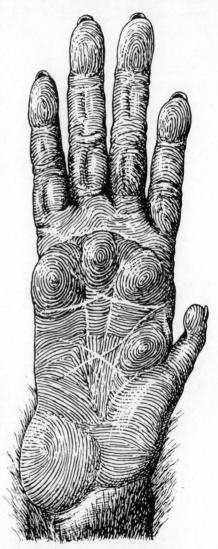

Figure 25 Hand of an adult langur showing the feeble thumb, the dermatoglyphics and pads. The first interdigital pad has shifted proximally with the thumb.

60

and thus do not participate in gripping. In various platyrrhines the thumb and index finger are spread apart from the other digits in grasping small objects or branches, as has frequently been observed and as is shown by the example in figure 26.

The first digit of the foot has not become eliminated in any primate, but it has clearly degenerated in orang-utans among which more than half the individuals lack terminal phalanges and nails on their 'great toes'. With this single exception the first toes are generally somewhat more movable and stronger than the first fingers of the same species of nonhuman primates. Only in man and to a lesser degree in the eastern gorilla has the great toe lost nearly all its original grasping ability, partly because its separation from the second toe is limited to the short distance in front of the sole (plate 5). In general the first toes are better suited for firm holds in climbing than are the thumbs, since the former can often be spread and even opposed more freely and are then also supplied with more powerful musculature. Indeed, in many prosimians, all monkeys and most apes the first toes are so thumb-like that these primates used to be called Quadrumana or four-handed ones, to distinguish them sharply from man, who was then regarded as the exclusive representative of the Bimana.

As shown by the examples in the figures 23 and 24, the second digits have become much reduced in length and functional importance in various primates. They are comparatively feeble in most colobine monkeys and very much shortened in the aye-aye, slow loris, potto and angwantibo. The second finger of the last-named genus has become even completely eliminated on the outer hand with merely a rudimentary basal phalanx left hidden in the palm. In many prosimians and some platyrrhines the fourth toes and fingers surpass the middle ones in length and thus provide the longest digital spread when opposed to the first digits. The foot of man supplies still another example of the manifold evolutionary reductions in phalanges in response to special modes of locomotion. The phalangeal parts of the human toes II to V have become greatly shortened as an adaptation to terrestrial life, in particular the middle phalanges, which are completely lacking in the fifth toes of many human beings. Further modifications in the hands and feet of primates, also bearing on differences in locomotion and posture, will be discussed in connection with the specialisations of the skeleton and the bodily changes during growth.

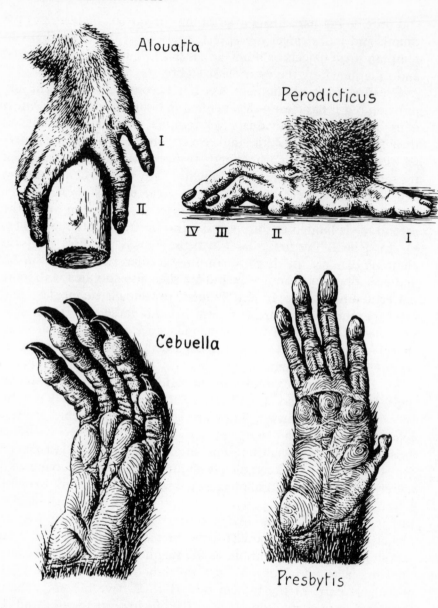

Figure 26   Right hands of a howler monkey, a potto, a marmoset and a langur
(after photographs). The howler monkey grasps the support between the fingers
II and III, the potto shows the extreme abductability of the thumb and the
vestige of finger II, the langur shows the rotation and proximal migration of the
small thumb in contrast to the lack of these conditions in the thumb of the
marmoset.

62

In this description of the role of the limbs in the predominantly arboreal life of most primates it must be mentioned also that no monkey or ape can climb an anywhere nearly perpendicular, smooth and branchless tree-trunk too thick to be securely grasped with its digits or even to be hugged between its arms. The only exceptions are the small marmosets which can hold their light weight adequately with their sharp claws, much as squirrels can. All others can ascend an upright column only by reaching around with their arms for pulling and by pushing with their legs from holds gained somehow by their feet, pressed against the bark of the tree, much as human natives can still climb trees of moderate thickness.

There is no doubt that the possession of a long tail can greatly influence posture and locomotion. This is most clearly apparent in the prehensile-tailed New World monkeys, the howler, spider and woolly monkeys, whose tails act much like fifth limbs. These most extremely specialised tails are not only very long and strong, but have developed a highly sensitised cutaneous zone on the ventral side of the distal half which bears, instead of the usual hair, skin ridges comparable to those on the palms and soles (figure 50). With these tactile, very mobile and exceptionally flexible tails the animals constantly explore their surroundings while climbing and can secure such firm holds that they are capable of freeing their hands and feet for other actions while hanging suspended head down by their tails alone. The grip of these tails is so strong that even when shot the animals often will not drop, but remain hanging by their tails alone for an amazingly long time. These prehensile tails are not only longer than the limbs, but have also a wider range of motion with an excellent sense of position, quite independent of visual control. These enviable structures are in constant use in balancing during locomotion or in securing the animal at rest as well as for obtaining food beyond the reach of the limbs.

In some primates, such as sifakas, tarsiers, capuchin and spider monkeys, the long tails often act as third legs while the animals stand upright like tripods. The extremely long and specialised tails of tarsiers also form a very effective support by being pressed against vertical twigs on which these small creatures sleep in an upright position during daylight. In the very large variety of pro-simians and monkeys with long, but not prehensile, tails, these appendages serve most likely as balancing rods in quadrupedal running on swaying branches and for changing the direction of the trunk in leaping by means of rapid bending of the tail. In nearly all New World monkeys the tails are relatively thicker and heavier than they are in Old World monkeys and hence can be particularly effective in balancing. The weight of the tail vertebrae equals from about fifty-five to eighty-five per cent of the weight of the precaudal vertebrae in the former, whereas only between about ten to fifty-two per cent in the latter, according to data obtained by the writer (1962). The long tail of the aye-aye bears such a thick coat of hair that it can cover the entire body when folded over during rest. Other species, particularly of *Pithecia* and *Colobus,* also possess unusually bushy tails which must be of some advantage.

The proportionately short tails of many macaques and baboons can, of course, have no influence on locomotion, but seem to play a minor role as means of communication, much as in dogs, by their changing direction and motions. Some of these short tails may also be useful as fly-whisks for the genito-anal openings, corresponding to their function in ungulates. In certain species, such as pig-tailed macaques and mandrills, these short tails are carried at least partly upright and thereby help to prevent the young from falling off while riding on the mother's hips (figure 22). The tail of *Cheirogaleus* is distinguished by its capacity for storing fat during times of plenty until it carries with it a greatly distended appendage, similar to that of the North African fat-tailed sheep.

There can be no doubt that among primates the reduction of the tail represents a specialisation which developed repeatedly and independently among all major groups. In the prosimians all Tupaiiformes have very large tails and so have most Lemuriformes, except the subfossil *Megaladapis* and the largest living representative, the indri. Of the superfamily of Lorisiformes the Galagidae all have well developed tails, but in every living species of the family Lorisidae the tail has degenerated greatly, being reduced to a useless stump, containing only few vertebrae. As is shown by the exact data in Table 1, the tails of New World monkeys are usually two and more times longer than the trunk (measured from suprasternal notch to upper end of pubic symphysis) and contain from twenty-three to thirty-two vertebrae, but in the uakaris (*Cacajao*) the tail has become much shorter than the trunk and has only about thirteen segments left. Among catarrhine primates all monkeys have long tails with well over twenty vertebrae, except for several species of macaques and baboons, as well as '*Cynopithecus*' monkeys, in which

Table 1 Averages of the length of the outer tail in percentage of the length of the trunk in adult primates and the average numbers of tail vertebrae in the same genera according to the author's publications of 1956 and 1961 (with a few new data).

Genus	Length	Vertebrae	Genus	Length	Vertebrae
Tupaia	191	24	Saimiri	207	28
Lemur	202	26	Cebus	196	23
Microcebus	204	25	Lagothrix	264	26
Avahi	167	23	Ateles	248	31
Propithecus	188	25			
Indri	16	10	Macaca irus	190	27
Daubentonia	230	23	Macaca mulatta	70	17
Loris	0	7	Macaca maura	11	6
Nycticebus	9	8	Macaca sylvana	1	3
Perodicticus	18	11	Papio hamadryas	133	23
Galago	248	25	Papio sphinx	17	9
Tarsius	307	27	Cercocebus	173	25
			Cercopithecus	178	27
Leontocebus	239	32	Erythrocebus	173	24
Callithrix	224	29	Presbytis	199	28
Aotus	197	25	Nasalis	154	25
Pithecia	176	24	Colobus	207	26
Cacajao	65	13			
Alouatta	221	27	Hominoidea	0	2–4

the tail has 'suddenly' been reduced in varying, marked degrees, including the nearly complete loss of an outer tail in the so-called Barbary Ape. In the entire subfamily of Colobinae only the isolated single species of *Simias* possesses an exceptionally short tail. Finally, gibbons, great apes and man share the total disappearance of an outer tail after embryonic life and the extreme reduction of caudal vertebrae to few vestiges. From this survey it is very evident that the loss of a tail is not necessarily connected with a terrestrial mode of life, nor with brachiation, as has occasionally been claimed. Tails are equally well developed in the terrestrial patas monkeys as in the highly arboreal guenons or langurs and the baboons and geladas of the Ethiopian rocky highlands possess very much longer tails than the mandrills and drills of the dense West African forests, whose caudal stumps, stuck in a coquettish curve above the hips, can at best have a decorative value.

The ends of long tails are much farther away from the heart than even the feet and it has repeatedly been found that captive monkeys, kept outdoors during cold weather, had frozen tips to their tails and no other bad effects. It seems tempting, therefore, to assume that selection in the rigorous environment of the northernmost species of macaques might have favoured individuals with reduced tails. On the other hand, in tropical Celebes there live other monkeys which are also stump-tailed. Furthermore, some species of *Colobus* and of *Rhinopithecus* thrive in regions with many intensely cold nights, yet have long tails with twenty-five to twenty-eight vertebrae. In conclusion it can merely be stated that the long and strong tails of the great majority of primate species do have a definite function in locomotion, but that the greatly reduced tails of a wide variety of species must have been retained for other reasons and that the nearly or entirely completed loss of an outer tail in a few prosimians, several macaques and all hominoids has evolved without any apparent relation to locomotion or posture.

The construction and shape of the trunk are closely connected with habitual movements and preferred postures. In lower primates the trunk is generally much more slender than in the higher species and the pelvis is farther removed from the thorax in direct relation to a longer lumbar region. This is strikingly evident from the great differences in the trunk skeletons of the macaque and the chimpanzee in figure 27. With the long lumbar region of monkeys the trunk is far more flexible than in apes in which the last ribs nearly

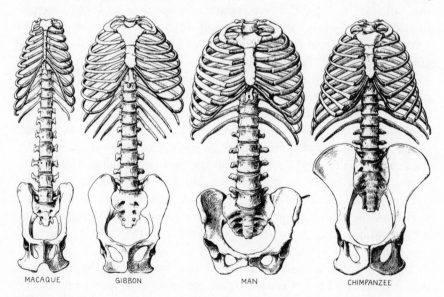

MACAQUE   GIBBON   MAN   CHIMPANZEE

Figure 27   Front views of trunk skeletons of four adult catarrhines, reduced to same total length (after Schultz, 1956).

touch the hip bones and thus permit hardly any lateral bending of the trunk during locomotion. With their extreme reduction of the lumbar region in length and number of vertebrae the great apes cannot compare with the lower primates in the ability to flex and twist the spinal column and thereby the entire trunk while climbing or leaping. When drinking water at ground level a monkey can remain seated and merely curve its spine in the numerous intervertebral joints of its lumbar region, but an ape has to tilt its comparatively inflexible and stout trunk in the hip joints, or else lie flat on its belly, as shown in figure 28. In typical quadrupedal monkeys the thoracic and abdominal viscera all hang underneath the vertebral column and the trunk is usually deeper than it is wide throughout its length. In the hominoids, on the other hand, the spinal column has migrated ventrally into the trunk cavities, which are wider than they are deep, so that the vertebrae really form a column near the centre of gravity when the trunk is held erect in climbing or bipedal posture. This nearly central position of the vertebral bodies in cross sections of the chests of apes and man has also been contributed to by the migration of the shoulder blades from the sides of the narrow thorax of monkeys to the broad back of all hominoids. These conditions will be discussed more fully in the chapter on growth, since they undergo significant changes during individual development.

All primates have necks of moderate length and flexibility, sufficient for moving the head without shifting the trunk in such close dependency as, e.g., in elephants. In quadrupedal posture the cervical spine of primates is commonly bent upward to bring the head in position for looking toward or above the horizon, since the sagittal movability at the cranio-vertebral joint is quite limited. If the latter joint lies fully at the rear end of the skull, as is especially found in prosimians, very little dorsoflexion is needed in the neck, but if the occipital condyles have shifted to a position underneath the skull, as in many monkeys and most of all in man, the head would naturally look downward with the neck held horizontally. Rotation of the head takes place mostly in the first intervertebral joint and involves the other cervical vertebrae to only a limited extent. In most primates it reaches at best to only about ninety degrees to either side, but tarsiers are distinguished by their uncanny ability to rotate their heads much farther until they seem to look straight back without having moved their trunks in the least. In

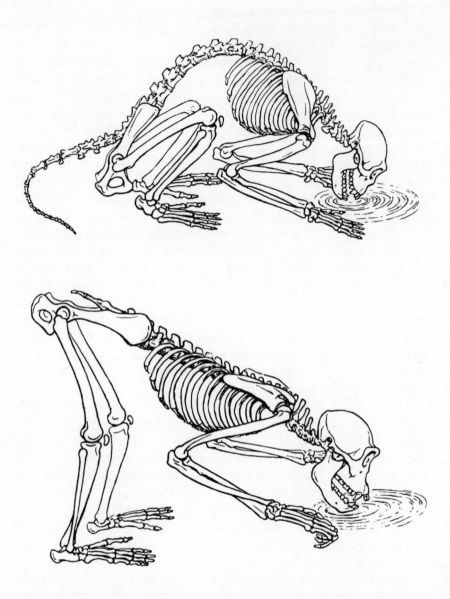

Figure 28 To bring the head to the ground, as in drinking, the monkey can remain sitting due to the flexibility of its spinal column in particularly the long lumbar region, but the ape must either crouch down or tilt the body in the hip joints on account of its few lumbar vertebrae, partly wedged between the long hip bones. Note also the different positions of the shoulder blades and shoulder joints.

adult great apes the neck becomes almost unrecognisable as such, since it is hidden in front by the huge face, which reaches nearly to the suprasternal notch. On both sides the neck is sunk between the high shoulders and in the back it is covered by the powerful nuchal musculature which stretches from high up on the heavy head across to the shoulders. Marked dorsoflexion of the cervical region of the vertebral column is prevented by the exceptionally long dorsal spines in the necks of all great apes. What the latter lack in the movability of their bulky trunks and thick necks, they can easily compensate for in the dexterity of their limbs which permit a greater range of motions than in most monkeys and prosimians. The shoulder and hip joints of hominoids have undergone certain definite changes in construction which, though inconspicuous, do facilitate the movements of the arms and legs in every direction. In lower primates the exact formation of these joints is mostly adapted for sagittal movements of the limbs, rather than for lateral ones. In some prosimians the upper arms and thighs are also more extensively enclosed under the skin of the trunk than in any Simiae, whereby the abductability of the limbs is limited. *Microcebus,* for example, has skin-folds reaching from the sides of the trunk nearly to the elbows and knees.

In this chapter on locomotion it must also be mentioned briefly that the speed of movements differs widely among some primates. The treeshrews, most lemurs and all bushbabies are very quick and agile in their movements and can run, climb and leap with astounding dexterity. The opposite extreme of the prosimians is characteristic of the Lorisidae which move with deliberate sluggishness in stalking their prey and in choosing holds among the branches in their forest homes. In climbing the lorisids rarely release more than one limb at a time while grimly maintaining their grip on the supporting branch with the other three. Only the heads and jaws can move with lightning rapidity, as the writer witnessed when an old potto bit a man offering food deep in the hand. Extremely rapid and precise movements have been observed in bushbabies which can leap for and catch flies in mid-air and in gibbons, known to grab birds on the wing. Of the two Asiatic ape families the hylobatids are as quick in every motion as the orang-utans are slow and unwilling to hurry. The author has often surprised capuchin monkeys and howler monkeys feeding in the same trees, but the former will have climbed and jumped clear out of sight long before the latter have reached some higher branch on their original tree. Most Old World monkeys can move faster than any man, and the terrestrial patas monkeys, living in savannas, have been reported to outrun a car going at thirty-five miles an hour and, incidentally, not trying to seek safety in a tree, as most other monkeys would, but relying on their great speed. The latter is due to the slender body-build and long limbs of this remarkable monkey, in which it resembles the distinctive body proportions of the swift cheetahs among the Felidae.

Lemurs, bushbabies, capuchin monkeys, guenons, langurs, etc. can leap from limb to limb and tree to tree for incredible distances without coming to any harm most of the time. They seem to prefer landing on slender peripheral twigs of which they grasp many with both hands and feet, bending them with their impact and using the springy rebound often for an immediate further leap. In jumping onto a firm branch or in dropping to the ground the shock is absorbed by the combined effect of many devices in the bodies of typical quadrupedal primates. To begin with, the extremities become flexed in their main segments, i.e., not only in the elbows and knees, but also in the shoulders and hips. Such strongly flexed limbs are characteristic also for the standing and running postures of most lower primates (figure 5), while in the higher forms the limbs are usually held considerably straighter in standing and walking on all fours (figure 21, the two adults on the left). The flexibility of the hands and feet contributes also to shock-absorbing during locomotion and the first contact with the supporting surface is usually made with the digits and only subsequently with the wrists and ankles, so that at least the metacarpal and metatarsal parts act as springy levers in breaking the impact of the body weight in motion when coming to rest on the ground. Last but not least, the very limited, but nevertheless significant movability of the shoulder-and hip-joints, relative to the spinal column, is of direct help in 'bouncing' the trunk between the limbs. The shoulder-blades catch any impact transmitted by the arms by moving against the chest until held by the clavicles. The hip-bones, being directly united with the vertebral column by strong ligaments, act nevertheless as a springy lever in transmitting the load on the spinal column to the hip-joints far behind the sacro-iliac connection in all nonhuman primates.

With the erect posture and bipedal locomotion of man the problem of shock-absorption has changed, since the legs are no longer flexed in standing and body weight is carried in the hip-joints alone, which lie perpendicularly beneath and very near to the sacro-iliac joints without any effective lever action as in the hip-bones of quadrupedal primates. To compensate for these changes, the spinal column of man has become markedly curved and thereby acts not as a rigid, but as a springy column together with all of its compressible intervertebral discs which form real cushions in upright posture (figure 29). The foot skeleton of man has acquired in its profound transformations a marked sagittal and transverse arch which also contributes considerably to shock-absorption. In standing and walking man's curvature of the spine becomes more accentuated and the arches in his feet less so, as is readily evident from a corresponding reduction in total body height and increase in foot length, demonstrable in every individual at the end of an active day. It seems, however, that the adaptations of the human body for bipedal locomotion have left it with decidedly less springiness than exists in most quadrupedal primates. No man could

Figure 29   The curvature of the vertebral
column and the relative size and position of
the pelvis in adults of ape and man (after
Schultz, 1957).

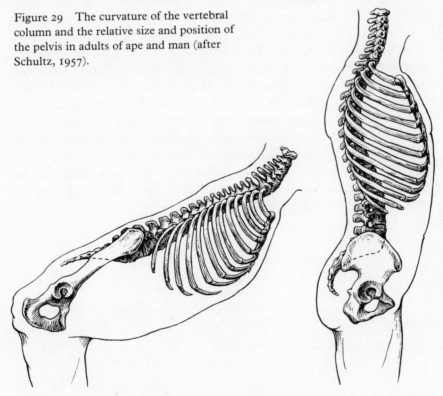

equal the old captive macaque which the writer himself saw escape and jump from the roof of a two-storied house to the paved street and immediately run away unharmed. Doctor Reynolds, who had studied wild chimpanzees in the Budongo forest, twice saw falls, due to the breaking of supporting branches, and states that, 'On the first occasion an adult fell about seventy feet; during the fall it held all four limbs stretched downwards and after landing in the under-growth it ran off. The second time a small chimpanzee fell about thirty feet.' From the occurrence of healed fractures in wild non-human primates, to be described later, it must be concluded that such accidents are not infrequently quite harmful.

Posture while resting is of great interest in view of the fact that normally all nonhuman primates spend more than half their lives in sleep and rest. As far as is known, primates can sleep sitting up or lying down on the dorsal, ventral or lateral sides of their trunks, with the last-named position being probably the most common one. In many prosimians the spine becomes strongly bent and the head and tail is tucked under during sleep. Pottos have been observed to sleep occasionally even suspended underneath branches. During cold nights monkeys seem to prefer sleeping in a sitting position, closely huddled together for warmth. When resting on branches, monkeys tend to hold on to some additional support with their hands and, in species possessing them, with their prehensile tails. The great apes, while resting in their nests, invariably grasp some nearby branches, even while sound asleep. Indeed, they rarely trust only one support, but will take hold of different branches whenever possible also during tree-climbing. While sitting for rest or station-ary feeding monkeys and apes generally hold their trunks nearly upright and thus are well prepared for the ease with which they can also assume an erect bipedal posture.

Chapter 6

Skeleton

THIS entire book deals most of all with the specialisations of primates inasmuch as the features which have become different in single species or in groups of them must be continually discussed in the chapters on locomotion, sense organs, growth, behaviour, etc. This chapter, therefore, is really a continuation of the story of primate differentiation in an attempt to show by means of further selected examples the manifold specialisations of anatomical systems which are of outstanding aid in determining the phylogenetic interrelationships between primates.

The primate skeleton has undergone a great many similarly directed or diverging changes during the past, which have been studied more intensively than the specialisations of the soft parts, not only because skeletons are available in collections in far larger numbers than are preserved bodies, but also on account of the direct comparability of the skeletons of recent primates with those of fossil ones.

In contrast to the skeletons of many other mammals, those of primates have in general remained fairly conservative in their phylogenetic changes. For instance, even all recent primates still possess well-functioning clavicles, at least the proximal halves of the fibulae, and hand- and foot-skeletons of primitive pentadactyl construction without the complete loss of any digital rays, no matter how far one or another may have become reduced in a few species. Inasmuch as primates are essentially arboreal and, particularly, because climbing requires great flexibility and freedom of limb movements, there have occurred no major fusions or reductions in the limb bones of primates. Unlike some exceptional other mammals, all primates have retained the ancient number of seven cervical vertebrae, while only their caudal ones have become greatly reduced in a widely scattered minority of species, as has happened also in a great variety of species belonging to other mammalian orders. Even the skulls of primates show nowhere near such extreme specialisations as have developed in whales, elephants and so many other animals, with grotesque tusks, horns and far-reaching changes in the entire dentition, which were all accompanied by intensive alterations in cranial construction.

The skeletal parts of the neck, trunk and tail of primates reflect in their number, relative size and topographic relations the varying shapes and functions of the corresponding body parts. As has already been mentioned, the cervical part of the spinal column normally always consists of seven segments, but these are proportionately somewhat longer in all hominoids, spider monkeys and tarsiers than in the other primates. The cervical portion of the spine is far more flexible than the adjoining thoracic portion and thereby enables the head to be moved with considerable independence of the trunk. In the great apes, however, the neck cannot be bent back very far on account of their extremely long dorsal spines which serve for attachment of the powerful nuchal musculature. In pottos, which are also distinguished by exceptionally long dorsal spines of the third cervical to first thoracic vertebrae, the bony tips of these spines are covered only by glabrous skin and actually reach above the surface, thus forming a defensive mechanism with the head and neck bent down.

The combined number of thoracic and lumbar vertebrae varies very extensively among primates, namely between fifteen and twenty-four according to the data in table 2. These numerical variations are of special interest since they show the extent to which the pelvic girdle has shifted along the spinal column. From an examination of the table it is evident that in all prosimians and in all monkeys of the New and Old World there are very rarely only seventeen and else at least eighteen segments in the spinal column of the trunk above the sacral vertebrae, which connect the hip bones with the spine. Among the prosimians it is the family of Lorisidae which is distinguished by the highest numbers of these vertebrae, indicating that in these primates the support of the spinal column by the hind legs *via* the pelvis has shifted caudally by several segments. The original number of thoracic plus lumbar vertebrae was in all likelihood nineteen and has been retained by such primitive species as most treeshrews, a few lemurs and also by various platyrrhine monkeys as well as the great majority of catarrhine monkeys. It is noteworthy that all the Old World monkeys are so uniform in this

Table 2 Ranges of variations and averages of the numbers of thoracic and lumbar and of sacral vertebrae in recent primates, totalling 1884 specimens (from Schultz, 1961).

Genus	Thoracic + Lumbar Vertebrae		Sacral Vertebrae		Genus	Thoracic + Lumbar Vertebrae		Sacral Vertebrae	
	Range	Average	Range	Average		Range	Average	Range	Average
Tupaia	18–20	19·1	3–4	3·1	Cacajao	19	19·0	4–5	4·5
Lemur	18–20	19·0	2–3	2·9	Alouatta	19–21	19·4	3–4	3·3
Hapalemur	18–20	18·9	3	3·0	Saimiri	20	20·0	3	3·0
Lepilemur	20–22	20·9	2–4	3·2	Cebus	18–20	19·6	3	3·0
Cheirogaleus	20	20·0	3	3·0	Lagothrix	17–19	18·2	3–4	3·4
Microcebus	19–20	19·9	2–4	2·9	Ateles	17–19	18·0	2–4	3·0
Avahi	19–20	19·9	3–4	3·3					
Propithecus	19–20	19·9	3–4	3·2	Macaca	18–20	19·0	2–4	3·0
Indri	20–21	20·5	3–4	3·5	Cynopithecus	18–19	18·8	3–4	3·6
Daubentonia	18–19	18·9	2–4	3·1	Papio	18–19	19·0	3–4	3·2
Loris	21–24	22·8	2–5	3·3	Theropithecus	19	19·0	3	3·0
Nycticebus	22–24	23·2	5–9	6·3	Cercocebus	18–19	18·8	2–4	2·9
Arctocebus	21–22	21·5	6–7	6·5	Cercopithecus	18–20	19·0	3–4	3·0
Perodicticus	20–23	21·8	5–9	6·6	Erythrocebus	18–19	18·7	3	3·0
Galago	19–20	19·3	3–4	3·2	Presbytis	18–20	19·0	2–4	3·0
Tarsius	18–19	18·8	2–3	2·9	Rhinopithecus	19	19·0	3–4	3·1
					Nasalis	18–19	19·0	3–4	3·1
Callithrix	18–20	18·9	2–3	2·9	Colobus	18–19	18·9	3	3·0
Leontocebus	19–20	19·0	2–3	2·9					
Callimico	19	19·0	3	3·0	Hylobates	17–19	18·2	3–6	4·6
Aotus	20–21	20·9	3–4	3·1	Symphalangus	15–18	17·3	4–6	4·7
Callicebus	19–20	19·7	3	3·0	Pongo	15–17	15·9	4–7	5·4
Pithecia	19	19·0	3	3·0	Pan	16–18	16·8	4–8	5·7
					Gorilla	16–18	16·6	4–8	5·7
					Homo	16–18	17·0	4–7	5·2

feature. That the hominoids are the only primates possessing the lowest of all the numbers of thoracic plus lumbar vertebrae is evident from the table and demonstrates the unique evolutionary migration of the pelvis in a cephalic direction by as much as four segments in some siamangs and orang-utans. This trend has not reached as high an extreme in man as in the great apes, of which the orang-utan has become most specialised in this respect. The number of movable segments in the trunk naturally influences its flexibility which is generally greater in the lower primates, with nineteen or more vertebrae between the shoulder and pelvic girdles and with long and slender bodies, than in hominoids with reduced numbers of intervertebral joints and proportionately short and stout trunks.

75

Since this flexibility is naturally always very much greater in the lumbar region than in the thoracic one, which is hindered in its movements by the rib cage, it is of interest to consider also the numbers of segments in each of these vertebral regions. These numbers are particularly variable intraspecifically because a last thoracic segment can change into a first lumbar one with comparative ease unilaterally or bilaterally whenever in early development a costal element happens to remain independent as a movable rib or fuses with the transverse process. Asymmetrical thoraco-lumbar transition vertebrae with a rib on one side and a more or less elongated transverse process on the other side are quite common in many primate species. These cases indicate an exceptional instability at the border between the two vertebral regions. The highest numbers of thoracic vertebrae have developed in the Lorisidae, namely from thirteen to seventeen. Among platyrrhines several genera of the higher forms also commonly have fourteen and occasionally fifteen chest segments, whereas in all catarrhine monkeys there are never more than thirteen and usually only twelve of these vertebrae. The same is true of man, but in the gibbons and African apes the corresponding numbers vary between twelve and fourteen and in orang-utans between eleven and thirteen. The number of lumbar vertebrae ranges among primates from three (exclusively found in apes) to ten (occurring only in the prosimian genus of *Lepilemur*). Six or seven lumbar vertebrae most likely represent the original, unspecialised condition, still persisting as the most frequent variations in the great majority of all lower primates. In only a few platyrrhine genera has the number of lumbar segments become reduced to four or five in either all, or at least some individuals, and these are the very types distinguished also by exceptionally large numbers of thoracic vertebrae. The hominoids are outstanding among all primates by having the shortest lumbar region, containing on an average only five vertebrae in gibbons and man, four in orang-utans and even only 3·6 in the African great apes. Not only in the number of segments, but also in its relative length has the lumbar region of the pongids been extremely reduced, since it measures less than one-fourth of the total length of the presacral spine in the latter, about one-third in man and more than four-tenths in the majority of monkeys. These striking differences in the relative length of the lumbar region are readily seen in a comparison between the trunk skeletons of the various primates in the

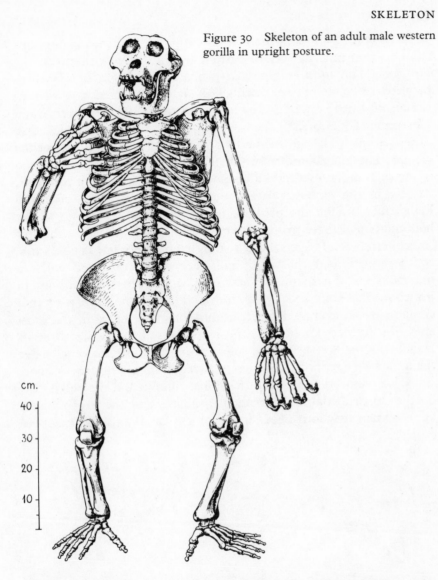

Figure 30   Skeleton of an adult male western gorilla in upright posture.

cm.
40
30
20
10

figures 5, 10, 17, 27, 29 and 30. In the great apes the lumbar region is usually wedged so far between the greatly lengthened hip-bones that only the upper few lumbar vertebrae can be freely flexed and the last pair of ribs approaches the pelvic crest extremely closely. It is easily understood, therefore, that there can be hardly any lateral movement within the stout and broad trunks of apes, in sharp contrast to the graceful bending of the lumbar region in the slender, narrow trunks of typical monkeys.

77

The sacral vertebrae, which become solidly fused during development to transmit the weight carried by the spinal column to the hip bones and hind legs, are in primates composed of anywhere between two and nine segments. As shown by table 2, a sacrum containing three vertebrae is by far the most frequent occurrence and represents undoubtedly the unspecialised condition, which still varies by only one segment up or down in the great majority of the genera. The inclusion of more than five vertebrae in the formation of the sacrum has evolved independently in two groups of primates, namely in the closely related genera *Nycticebus*, *Arctocebus* and *Perodicticus* among the prosimians and in the entire series of the hominoids. All these primates with exceptionally large numbers of sacral vertebrae are specialised also in having only few caudal ones, but a reduction of the tail is not necessarily combined with an increased sacral region, as shown by the short-tailed species of macaques, baboons and by the indris. Though the human sacrum contains on an average fewer vertebrae than the sacra of the great apes, the former has become more specialised in the width of its lateral wings which supply the contact with the hip bones and wedge them apart (figures 27 and 31). Bipedal man requires a more extensive connection between the spinal column and the pelvis than do the quadrupedal and bimanual primates and the exceptionally large human newborn needs a wide birth-canal and this has been

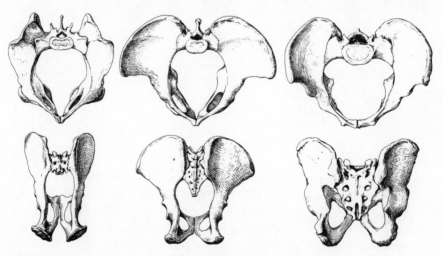

Figure 31   Cephalic and dorsal views of the pelves of adults of a macaque, a gorilla and a man, reduced to comparable sizes (after Schultz, 1961).

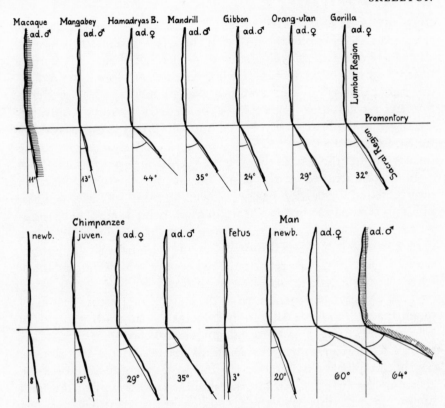

Figure 32   Exact ventral profiles of the lumbar and sacral regions of the vertebral column in some catarrhines, showing the differences in the formation of a promontory (after Schultz, 1961).

accomplished largely with the acquisition of extra wide and thick sacral wings. Due to the latter, the human sacrum weighs about one-fifth of the entire weight of the precaudal spine, whereas in all nonhuman primates this relative weight is considerably to very much smaller. Man's sacrum has become further specialised by being bent back abruptly, rather than following closely the general direction of the presacral spine, and by thus forming a clear promontory at the lumbo-sacral border (figure 29). This extreme innovation in the human vertebral column, which does not fully develop until long after birth, represents a vital adaptation for keeping the birth-canal open by tilting the ventral side of the solid adult sacrum until it is parallel to the pubic symphysis which it approaches more closely than in all other primates where the sacrum lies much higher above

79

the ventral part of the pelvic girdle. A sacrum which is bent dorsally from the general axis of the presacral spine is found also in adults of all man-like apes and even in some monkeys, but never is it bent nearly as far as in adult man in whom the sacrum forms the roof of the pelvic cavity in the erect posture (figure 32). This extreme human specialisation represents a mechanically risky acquisition inasmuch as the load carried by the lumbar vertebrae is precariously transmitted to the sacrum which stands at nearly right angles to the former, so that the last lumbar segment tends to slip ventrally, if not sufficiently anchored in the joints of the dorsal arches. Such an intervertebral dislocation has never been found in monkeys or apes, but is unfortunately not at all rare in man, being known as spondylolisthesis, a condition which can cause severe suffering and interfere with normal birth. Maladjustments of the vertebrae at the lumbo-sacral border have been reported as amazingly common in Eskimos, whose hard life calls for much heavy lifting.

It has already been mentioned in the chapter on posture that in all quadrupedal monkeys the thorax hangs underneath the spinal column and, being a partly cartilaginous and flexible basket, tends to be deeper than wide, whereas in all the hominoids the thoracic part of the spinal column moves during postnatal growth toward the centre of the broad chest and the shoulderblades shift to the back (figure 33). In this way the backbone really becomes a nearly central *column* around which the weight is distributed when man and apes are more or less upright. This condition has become most pronounced in man whose fully erect posture has also produced the comparatively thickest lumbar vertebrae, just as a tree trunk is thickest at its base while a bridge beam can have a more even diameter throughout its length. The various figures show also that the shoulder joints of monkeys lie in one frontal plane with the sternum, whereas those of the hominoids in one plane with the spinal column, a location which is better suited for brachiating than for quadrupedal locomotion and requires somewhat longer clavicles.

The hip-bones of the lower primates and particularly the parts in front of the hip-joints, the ilia, are comparatively slender, corresponding to their narrow trunks. In all the hominoids with their stout trunks the ilia have become very much wider and thus provide a much larger surface for the attachment of musculature (figures 10, 27, 30 and 31). The ilia of man appear especially broad through the unique enlargement of the sacral surface, needed for a firm connection between the sacrum and the hip-bone in a body standing on the hind limbs only. In addition the human ilia are highly specialised by having become much shortened as well as by having the sacral surfaces pushed close to the hip-joints (figure 34) and brought perpendicular above them, instead of far in front of them, as in nonhuman primates. As is best shown in figure 29, the body weight carried by the vertebral column is transferred from the sacro-iliac joints to the hip-joints by the long and springy levers of the ilia

Figure 33   Cephalic view of thorax and right shoulder girdle of an adult macaque and an adult man (after Schultz, 1957).

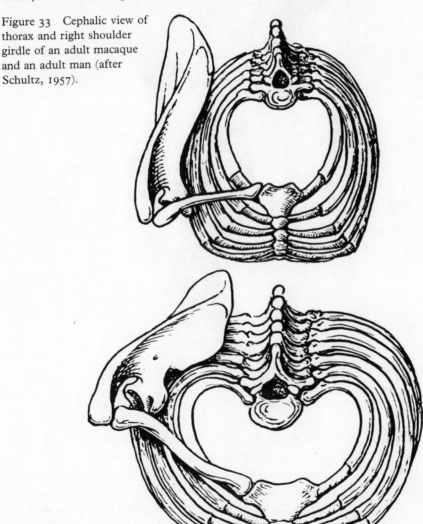

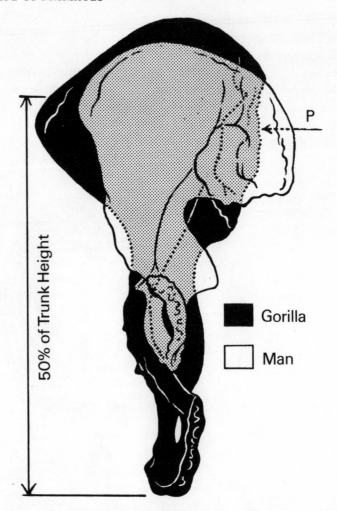

Figure 34   Medial views of right hip-bones of a gorilla and of a man, super-imposed and reduced to same trunk height, P = level of promontory. Note differences in size of sacral surface and in total pelvic height (after Schultz, 1936).

in quadrupedal primates, but by the short perpendicular parts of the ilia in bipedal man, a specialisation which prevents tilting of the pelvis in erect posture. In the hominoids, whose thorax approaches the pelvis very closely, the ilia naturally have to follow the direction of the last ribs with which they are connected by abdominal muscles. In the great apes, therefore, the ilia stand nearly in a frontal plane to conform to the lower end of the funnel-shaped, wide thorax, but in man they are directed closer to a sagittal plane, thus following the lower ribs of the human barrel-shaped chest (figures 27 and 30). All these many and very marked evolutionary transformations in the proportionate size and relative position of the human pelvis and its parts have been accompanied by intricate changes in the function of certain muscles, connected with the hip-bones, so that they can act most advantageously in upright locomotion. From the fortunate finds of several hip-bones of the fossil Australopithecines it has to be concluded that these primates were bipedal, because these bones show practically all the essential specialisations which are characteristic for man and which have been decisive in the unique evolution of the erect posture of all hominids.

The limb bones of primates have become comparatively little specialised, retaining most of the ancient designs for quadrupedal, arboreal locomotion with at least some capability for grasping with the pentadactyl extremities. Fusion of radius and ulna with the consequent loss of rotary movement in the forearm and hand, common in many fully terrestrial mammals, is unknown among primates. Reduction of the fibula and its fusion with the tibia, also quite common in other quadrupeds, has appeared only in tarsiers (figure 16) and some of their fossil relatives but not in other primates. Even though single digits have become atrophied in a great variety of prosimians, monkeys and hominoids, these losses are very modest compared with the far-reaching evolutionary specialisations in the hands and feet of many other mammals with different modes of locomotion. In the conservative order of primates the front limbs permit a wider range of motions than the hind limbs, being connected with the freely movable shoulder girdle instead of the fixed pelvic girdle and having much better rotating powers in the forearms than in the corresponding parts of the extremities at the rear. The latter serve mainly for pushing or propelling the body forward, while the front limbs act chiefly in pulling, holding or suspending it with the added ability of wide lateral movements, surpassing those of the hind legs as a general rule. The hands being not only under ready control of the eyes, but usually also more flexible than the feet, are used for a greater variety of purposes. While the hind limbs of all primates function almost exclusively for supporting and holding the body, the front limbs not only participate in carrying the body weight, but much of the time they are also free for tactile exploration, for grasping food and conveying it to the mouth, for carrying the young or any objects and for all the further detailed manipulations of which the hands of primates are capable.

The long bones of the limbs have developed some widely differing proportionate lengths among primates according to their specific modes of locomotion, as has already been partly discussed in the preceding chapter. In the quadrupedally running and climbing species the forelimbs are usually somewhat shorter than the hind limbs, but not nearly as much so as in the leaping forms which rely chiefly on their hind legs. This is best shown by the intermembral index, which expresses the total length of humerus and radius as a percentage of the total length of femur and tibia, and averages among the latter types only 54 in tarsiers, 58 in sifakas and 67 in bushbabies, whereas among the many former types anywhere between 70 and 100. The same index rises to averages above 100 only in the brachiating spider monkeys and in all of the man-like apes, among which the best brachiators have by far the highest intermembral indices with averages of 132 in gibbons, 144 in orang-utans and 149 in siamangs. The last and most extreme value equals almost three times the lowest of all averages, found in tarsiers (compare figures 16 and 17). The mean index of about 71 for adult man falls within the relevant range for the majority of prosimians and monkeys and permits the conclusion that the relation in length of the upper and lower extremities has not undergone nearly as much change in bipedal man as in the leaping or in the brachiating primates. This is due to the fact that not only the legs, but also the arms of man are longer in relation to the length of the trunk than in most monkeys. Indeed, the arms of man equal in this relative length those of many chimpanzees and gorillas, as shown by the examples in plate 7, and the length-ratio of leg and trunk, though very high in adult man, is surpassed by that of tarsiers and equalled by occasional individuals of various other primates.

In their relative thickness the long bones of the limbs also vary between extremes which are remarkably far apart and this naturally and chiefly on account of the great differences in the loads entrusted to them. Marked changes in the relative length of the limbs, how-ever, do not affect the thickness of the shafts of the limb bones, so that the latter can appear to be much more slender or stout than expected on the basis of the body weight of the particular species. This is clearly illustrated by the examples in figure 35, which compares the slender thigh-bone of a small sifaka with the robust

84

1 Old ivory carvings of Japanese macaques, one with toads and the other with an octopus under a basket (author's collection).

2 The young chimpanzee of E. Tyson, 1699, standing erect with the aid of a cane.

3 Macaque as caricature of man. Engraving by T. Landseer in a volume of similar pictures, entitled *Monkeyana* (London, 1827).

4 Satirical French engraving of the 18th century showing clothed macaques in human attitudes.

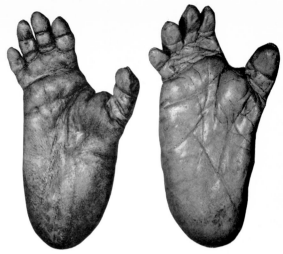

5 Feet of adult male gorillas, A=western, B=eastern subspecies (after Schultz, 1931).

6 Hands (above) and feet (below) of a macaque (left) and a man (right) at corresponding early foetal stages of development (after Schultz, 1957).

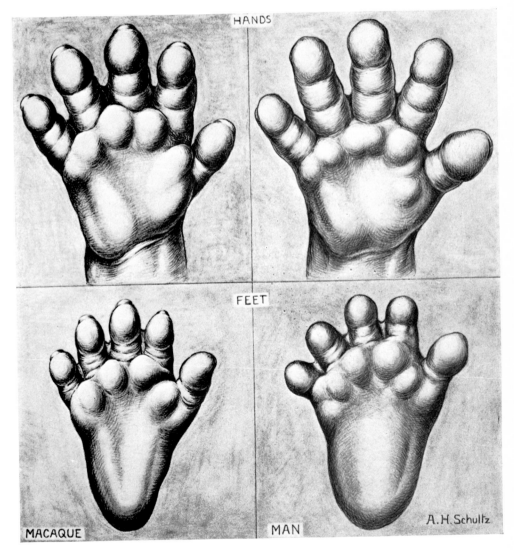

HANDS

FEET

MACAQUE

MAN

A. H. Schultz

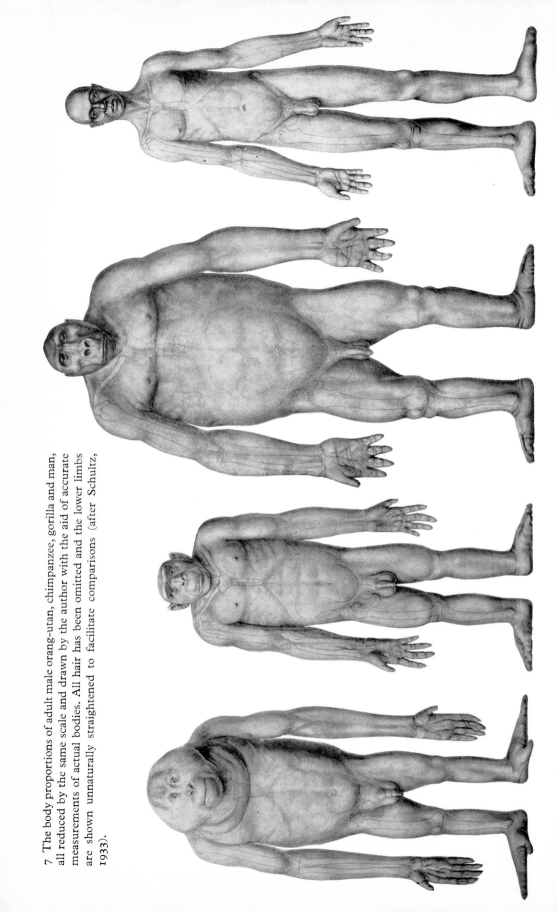

7 The body proportions of adult male orang-utan, chimpanzee, gorilla and man, all reduced by the same scale and drawn by the author with the aid of accurate measurements of actual bodies. All hair has been omitted and the lower limbs are shown unnaturally straightened to facilitate comparisons (after Schultz, 1933).

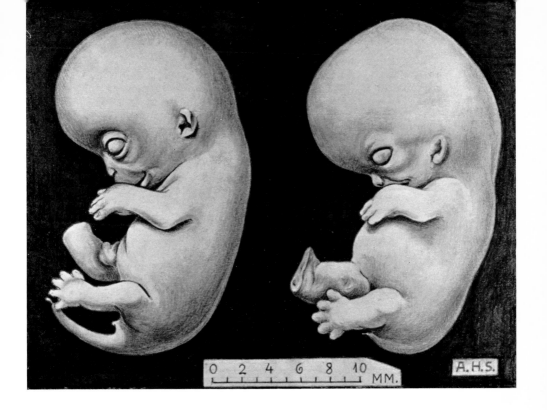

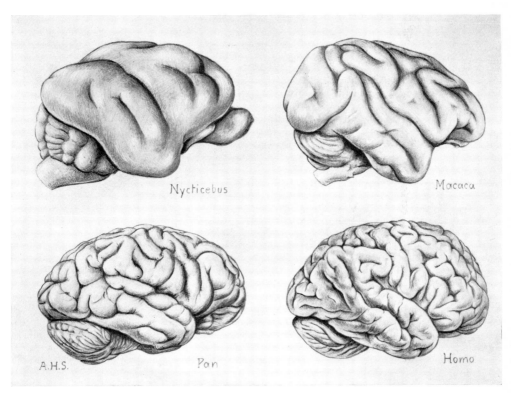

8 Early foetuses of macaque (left) and man of the same size and nearly the same age (macaque has known conception age of 44 days and man an estimated age of 49 days). Drawn from photographs of specimens in the collection of the Carnegie Laboratory of Embryology.

9 Right sides of brain of four adult primates, reduced to the same length (after photographs).

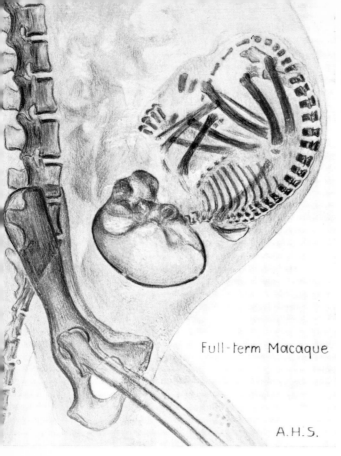

Full-term Macaque

A.H.S.

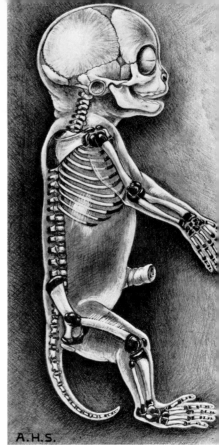

A.H.S.

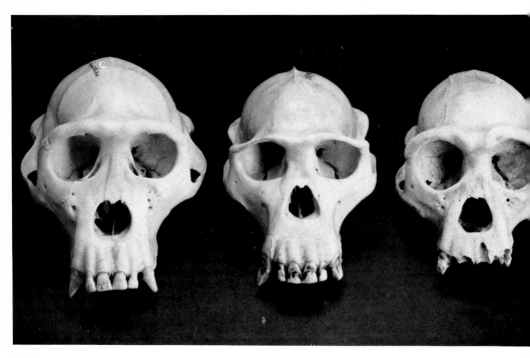

12 Skulls of adult male West African chimpanzees showing many marked individual variations (after Schultz, 1963).

10 The relation in size between a macaque foetus at full term and the maternal pelvis (after an X-ray photograph of a normal female rhesus monkey at the end of its pregnancy).

11 Macaque foetus from the middle of prenatal life showing the incomplete ossification in skeletal development. The specimen has been chemically stained and cleared until the cartilage has become deep blue, the bone white and all other tissue transparent.

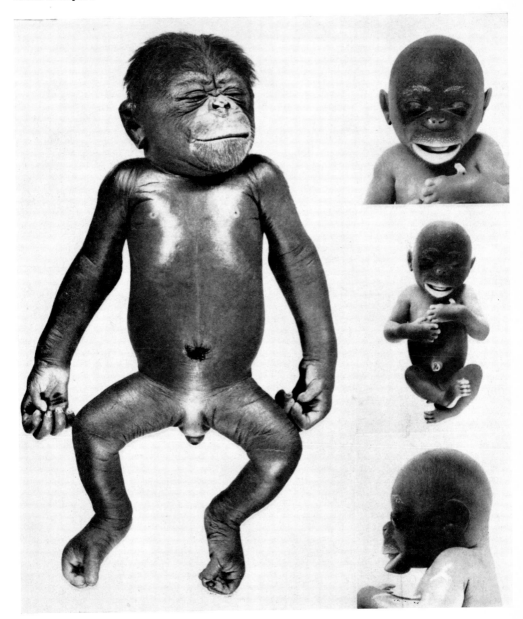

13 Chimpanzee foetuses of the fifth (right) and seventh month (after Schultz, 1940).

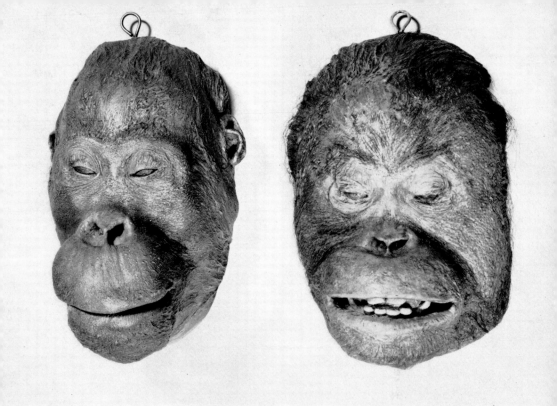

14 Death masks of two adult female North-Bornean orang-utans showing very marked individual differences (after Schultz, 1950).

15 Head of an adult *Australopithecus africanus*, reconstructed on the cast of a skull by the writer.

16 Head of an adult female *Homo erectus* from China, reconstructed on the cast of a skull by the writer.

one of a giant *Megaladapis*. The striking difference in relative
thickness between these bones of similar total length is only partly
due to the great discrepancy in body weight and is accentuated by
the fact that the leaping prosimian *Propithecus* has greatly lengthened
its lower extremities, whereas the extinct *Megaladapis* with its
puzzling locomotion has by far the shortest relative length of hind
limbs among all primates (figure 16).

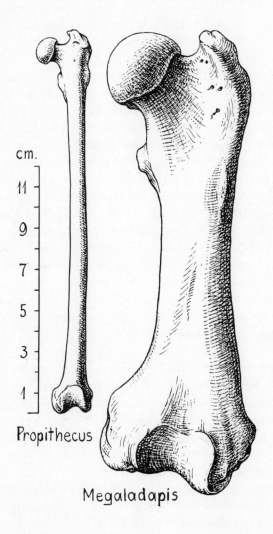

Figure 35   Left femur of an adult *Propithecus* and of an adult *Megaladapis*,
both reduced to same scale.

In relation to body weight orang-utans have remarkably thin and light long bones of the lower limbs in accordance with the comparatively weak development of their legs, except for their greatly lengthened feet. In a similar way and by considering the size of the body, man is distinguished by his slender and light arm bones and, particularly, by the low relative weight of his hand skeleton, showing the effect of the loss of locomotor functions in the human arms.

It must be emphasised again that in the life of primates the hands play a very prominent role by serving for remarkably many-sided purposes. For this reason the hands have undergone numerous adaptive modifications while retaining the essentials of the original anatomical groundplan, as is evident most of all in the conservative skeletal construction. The proportions and other details of the outer form of the adult hand are the features which appear to be particularly diversified, as shown by the examples selected for figure 24. In the primitive Tupaiiformes the thumb is as yet incapable of effective grasping, being merely abductable and not truly opposable. Handholds are instead accomplished by means of simple digital flexion and by the sharp, curved claws which are fully adequate for holding the light bodies of these small animals on the rough surface of trees or bushes. The same can be said of the hands of the small Callithricidae which are also equipped with claws (except on the first toes) and have thumbs with no real opposability (figure 26). The aye-aye is another form of recent primate with claws, but being of considerable size it also has a more freely movable thumb, suitable for holding in opposition to the long lateral fingers. The overwhelming majority of primates are characterised by the possession of more or less curved nails, which are probably retrogressive derivatives of the anatomically more elaborate claws and function primarily as mechanical supports for the digital touchpads. That nails, like claws, are useful also in scratching and digging needs no further comment, but it is of interest that on some fingers and toes of not a few species the nails have degenerated to mere vestiges and occasionally even disappeared completely, as on the first toes of orang-utans or the fifth toes of man. An extreme specialisation has been acquired by aye-ayes in their middle fingers which have become excessively slender even in their metacarpal part and, being armed with an equally thin, sharp claw, can be inserted into bark-holes for piercing and withdrawing the larvae of insects. As shown by figure 24, in a majority of prosimians the fourth finger is

86

the longest and the second finger very short. This condition has become especially pronounced in the Lorisiformes, the extreme reduction of finger II having developed in *Perodicticus* and *Arctocebus*. In the latter there is usually no outer trace left of an index finger, but a remnant of the basal phalanx is still embedded in the palm. This reduction of the second finger, together with the increased length of the fourth finger and the highly abductable thumb, produce the longest possible span for grasping branches (figure 26).

The hands of platyrrhines are characterised by thumbs which are not rotated to face the other digits and which branch from the palm near the base of the index finger, so that they are not well adapted for grasping. Most likely for this reason it has often been noted that these monkeys tend to hold objects between the second and third fingers, rather than between the first and second, as is normal for all catarrhines (figure 26). The more or less complete loss of an external thumb in spider monkeys and the marked lengthening of their other fingers has changed the hands of these expert brachiators into mere grappling hooks. The same specialisation has appeared again in guerezas among the Old World monkeys. In all other catarrhines the thumb has become rotated so that its transverse axis forms an angle of somewhere near 90° with the transverse axis of the other fingers and at the same time the free thumb has shifted its place of branching from the palm toward the wrist (figures 24, 25 and 26). These changes, which are accompanied also by improved ranges of motion in all the joints of the thumb, chiefly facilitate the entire repertoire of finer manipulations, rather than the simple gripping of branches. The free, phalangeal part of the thumb is comparatively short in all Old World monkeys and great apes in contrast to that of man, whose thumb reaches to nearly the first interphalangeal joint of the second finger. The human thumb, however, has not become lengthened, as often claimed, but it is the total hand length (to tip of middle finger) which has become relatively much shorter in man than in any other recent hominoid. In percentage of the trunk length this total hand length averages only 37 in adult man, but between 40 and 53 in adult great apes and even 59 in adult gibbons. In the latter, as well as in siamangs, the free thumb has been highly specialised by having most of its metacarpal part freed from the palm, thus giving it a wide range of motion in addition to its extreme relative length. A corresponding specialisation, which is also unique among simian primates, exists as well in the first toes of all Hylobatidae (figure 23).

The proportionately broadest of all primate hands are found in adult gorillas in spite of their being commonly classified among the brachiators, which supposedly have long and slender hands.

The specialisations of the feet of primates are in many details remarkably similar to those of the hands. For instance, in most prosimians the fourth toes surpass the others in length and in the lorisiform types the strong first toes can be abducted to an extreme degree, while the second toes often are very much reduced. In all monkeys the first toes are proportionately short and slender, resembling the conditions of the thumbs. No primate feet, however, have such degenerate first digits as have appeared in the hands of several unrelated monkeys. Only in the orang-utan has the 'great' toe become so far reduced that it no longer contains the normal two phalanges in over half the individuals (figures 36 and 37). That the outer form of the foot of adult man has induced former scholars to separate man under the term *Bimanus* from all the quadrumanous primates, then known, is understandable since the human toes have lost all resemblance to hand-like digits. The specialisations of the feet of bipedal man do not chiefly consist of adduction and apparently increased length of his great toes, but in the extreme reduction of the other toes, which no longer have to serve in reaching around branches and have adapted themselves to terrestrial life through excessive shortening of the phalanges, particularly the middle ones (figure 37). In its relation to the length of the trunk as well as to that of the tarsal and metatarsal parts of the foot the first toe is actually shorter in man than in chimpanzees and especially gibbons. That the great toe of man had once been freely movable even in its metatarsal part is evident from the fact that the joint at its base is still somewhat concave and directed slightly sideways, indeed in some individuals as much as in some eastern gorillas, whose feet are in many respects most similar to those of man. All gorillas can abduct their first toes to a considerable extent, but without rotating them in the least when placed on a flat support. Even though toes II to V of gorillas appear outwardly to be about as short as in man, their phalanges have not become markedly reduced. For instance, the phalangeal length of the middle toe forms on an average thirty-two per cent of the total foot length in gorillas, but only nineteen per cent in adult human beings. In early

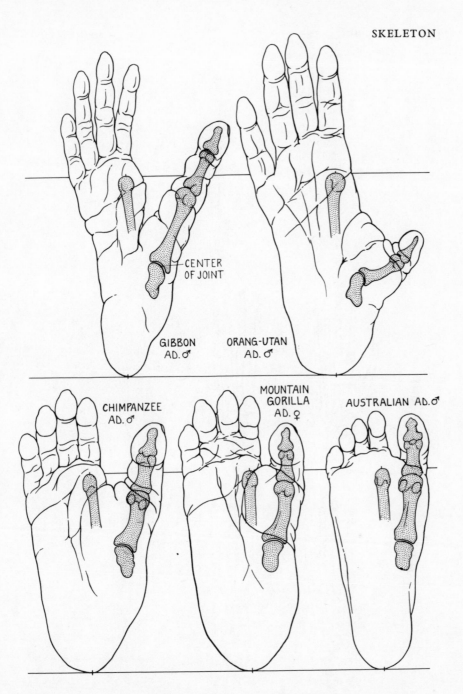

CENTER OF JOINT

GIBBON
AD. ♂

ORANG-UTAN
AD. ♂

CHIMPANZEE
AD. ♂

MOUNTAIN
GORILLA
AD. ♀

AUSTRALIAN AD. ♂

Figure 36   Feet of adult higher primates, showing some skeletal parts in their exact relation to the outer form, reduced to same distance from heel to second metatarso-phalangeal joint (after Schultz, 1950).

89

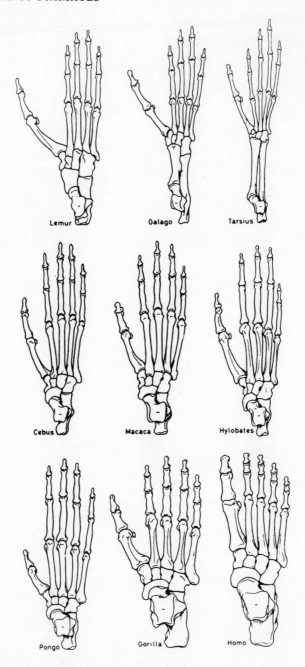

Figure 37   Right foot skeletons of some adult primates, all reduced to same length to tip of middle toe (after Schultz, 1963).

foetuses of man this percentage still equals thirty-five. With precise measurements on a large series of primate foot skeletons the author (1963) could demonstrate that the phalanges of the toes have generally become significantly shorter in such terrestrial species as geladas and patas monkeys and decidedly longer in such highly arboreal forms as most Colobinae, spider monkeys, Hylobatidae and, most of all, orang-utans.

Among the specialisations in the tarsal part of the foot skeleton of primates its great elongation in several unrelated forms is of special interest by showing the close connection between form and function. As is partly shown in figure 37, tarsiers as well as bushbabies are distinguished by the possession of two extremely lengthened tarsal bones, the calcaneus and naviculare, whereby they have gained an effective lever for leaping. This specialisation, unique among all mammals in regard to the particular skeletal elements involved, must have been acquired by these prosimians through convergent evolution and has reached quite different degrees of perfection among the various species of *Galago*. By elongating the tarsal part of the foot, rather than the more distal metatarsal one, as in a great variety of other mammals, the grasping function of the toes remained unaffected in the primates. The human tarsus is proportionately larger than that of any other simian primate, undoubtedly in response to its having to bear the entire load of the body. As a percentage of the total length of the foot skeleton the tarsal length equals on an average in man 50, in gorilla 40, in orang-utan only 26 and in monkeys anywhere between 28 and 37. The heel-portion of the tarsus, projecting behind the ankle joint, represents the power arm of the two-armed foot lever on which the muscles of the calf act when lifting the body. This power arm is relatively longest in man and gorilla in contrast to all other recent primates, as indicated by the examples in figure 38. The proportionately shortest parts of the heelbone, extending behind the fulcrum of the foot at the ankle joint, exist in a few prosimians, notably the tarsiers. In these forms even a modest contraction of the calf musculature will push the long distal parts of the foot much farther down, and thereby propel the body higher than in species with comparatively longer power levers of the foot, which are consequently not nearly as efficient jumpers.

Still another peculiarity of the human foot consists in the development of a longitudinal as well as a transverse arch in the tarsal and

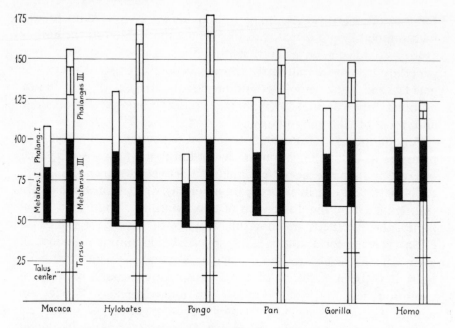

Figure 38 Diagrammatic representation of the average relative lengths of some of the main skeletal parts of the foot in macaques and hominoids, reduced to same tarsal + metatarsal III length (= 100) (after Schultz, 1963).

metatarsal parts. These arches are of considerable advantage in man's upright posture since they act as shock-absorbing springs, but tend to relax and flatten out when tired. The rotation of the human foot to a position in which the transverse axis of the sole stands at a right angle to the direction of the tibia, appears only gradually during early growth and still leaves slightly more pressure on the lateral than on the medial side of the foot, as is readily seen from the uneven wear of our shoe-soles. All these conditions support the conclusion that the foot of adult man has been made over from a primate foot designed for climbing and grasping. For the latter mode of action the foot of the orang-utan has become so highly specialised that it is placed almost entirely on its lateral side with the long toes strongly flexed during the rare and clumsy attempts at bipedal posture (figure 20).

Many special features of the primate skull are discussed in other chapters, wherever they seem to be most appropriate for illustrating age changes, sex differences, or conditions closely connected with sense organs and with the dentition. Of the great many other cranial specialisations, which have been described in an overwhelmingly extensive literature, only few selected examples can here be considered without getting involved in too much technical detail. The vertebrate skull serves to encase and protect the delicate brain and sense organs, to support the dentition and to provide a firm framework for the movements within the dental apparatus and those of the entire head. The size and shape of the skull in all its parts are, therefore, conditioned by the construction and purposes of the manifold organs within the head. The order of primates is characterised by more or less clearly recognisable general trends to enlarge the brain, to move the eyes toward the front, to reduce the nasal chamber as well as the masticatory apparatus, and to shift the direction and location of the cranio-vertebral joint from the rear toward the base. In the many mammals, which grasp their food directly with their lips and teeth, the face part of the head and skull lies entirely in front of the brain part and forms an almost straight continuation of the cranial base. This relation between the two main parts of the skull has been retained with little change in many fossil and recent prosimians, as shown by the skull of the lemur in figure 1. From the

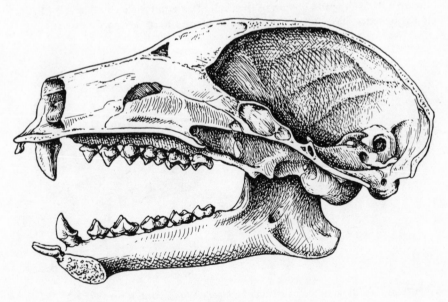

Figure 39   Midsagittal section of the skull of an adult lemur.

midsagittal section of another lemur skull in figure 39 it is seen that the axis of the face part is actually flexed toward the direction of the cranial base, on which the brain rests, but not nearly as much as in higher primates, such as the chimpanzee in figure 40. The latter

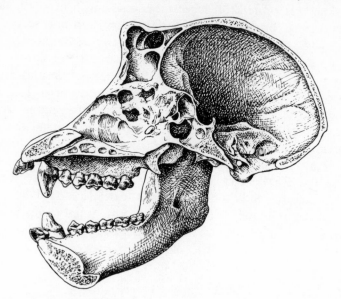

Figure 40   Midsagittal section of the skull of an adult male chimpanzee. Most of the turbinals have become lost and the lateral sinuses are partly exposed.

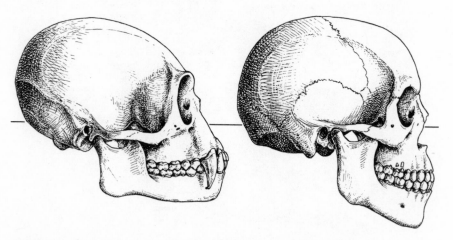

Figure 41   Adult skulls of a gibbon (*Hylobates lar* ♀) and a man (*Homo* ♂), reduced to same length of brain-case and posed in ear-eye horizon.

94

usually bring their food to the mouth with their hands and would benefit by a protruding dentition merely in the rare cases of defensive biting. With the gradual migration of the face underneath the braincase, closely connected with an increased flexure of the basicranial axis, the orbits become rotated toward the front and separated by a narrower interorbital region than in the skulls of the less specialised prosimians. This is very evident from a comparison between the side views of the skulls of prosimian and simian primates in the figures 1, 6, 7, 9, 41 and 42. In most of the former the orbits appear as wide ovals, but in the latter only as narrow slits, owing to the marked differences in the direction of the central orbital axis. In nearly all fossil and in all recent prosimians the eyes are protected on their lateral sides by a bony bridge, but in the rear the orbits are not separated from the temporal fossae by a bony partition, except incompletely in tarsiers (figure 2). In all simian primates such a partition has become greatly perfected, though in platyrrhines there remains a small gap of varying size, closed merely by a thin membrane (figure 9). It will be mentioned again later that the size of the orbits is greatly influenced by allometry so that such huge primates as *Megaladapis* (figure 6) have proportionately small

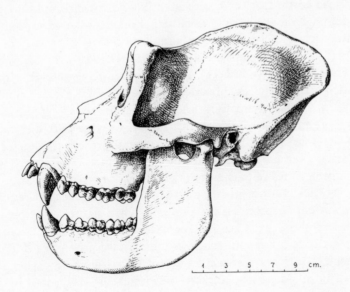

Figure 42   Skull of an adult male gorilla from Cameroon (after Schultz, 1964).

orbits and small ones such as lemurs (figure 1) or gibbons (figure 41), relatively large ones, forming a far more conspicuous part of the skull. In some of the nocturnal primates the orbits reach enormous sizes to accommodate their greatly enlarged eyeballs and thereby encroach on the nasal chamber till its upper part is reduced to a mere slit (figure 2). The upper opening of the lacrimal canal, through which the lacrimal secretion is normally drained into the nasal chamber, is situated outside of the orbital cavity in prosimians, but has shifted back to a position immediately behind the orbital rim in all simian primates, as is evident from all the examples of primate skulls in the accompanying figures. This is merely one of innumerable detailed cranial conditions which differentiate the two suborders of primates with significant consistency.

The relation in size between the brain part and the face part of the skull varies within very wide limits among primates according to the proportionate development chiefly of the brain and the dental apparatus. In many prosimians the large dentition demands a facial skull which greatly surpasses the moderate brain part in size (figures 1 and 6), whereas in most monkeys the face appears much smaller in comparison with the very noticeably enlarged brain-case (figure 9). The evolutionary reduction of the dental apparatus, and with it of the entire face, of recent man has not progressed to such a unique degree as is commonly assumed. For instance, in gibbons as well as in a considerable variety of small monkeys, especially of the New World, the facial part is no larger in relation to the brain part than in man (figures 9 and 41). Among simian primates the great apes and particularly the baboons have acquired exceptionally large and powerful dentitions and facial skeletons, especially in males with their huge canine teeth. This is only in part due to the factor of positive allometry. Early in postnatal growth these special-isations are as yet barely recognisable, since they appear only with the development of the permanent dentition. To serve the latter the masticatory musculature grows to corresponding sizes and requires large attachment surfaces on the posterior part of the mandible and on the sides of the brain case. The latter may not offer adequate surface for the temporal muscles, in which case a bony midsagittal crest develops after the completion of the dentition, when the muscles of both sides have come to meet each other without having as yet reached their final size. In old male gorillas such sagittal crests can attain surprisingly large dimensions all along the frontal and parietal bones, but they occur also in more moderate development among the other great apes as well as among many monkeys and prosimians with temporal muscles too large for the surface of the brain case. For similar reasons the nuchal musculature can induce the formation of high and strong, transverse, occipital crests whenever the part of the occiput facing the cervical spine is insufficiently large for powerful neck muscles which have to balance the head on the spinal column or hold it entirely in front of it. These occipital crests, therefore, are most common in primates in which the cranio-vertebral joint lies far to the rear, so that the centre of gravity of the head is way in front of the fulcrum and correspondingly thick muscles are required for action on the short power arm of the two-armed lever represented by the head. In adult man the centre of gravity of the head, posed in the ear–eye horizon, lies only about three centimetres in front of the occipital condyles, so that much less force is needed to balance the head than in any other adult primates. Conforming with the erect posture of man, the part of his occiput to which the dorsal neck muscles are attached is directed much more horizontally than in the apes or, particularly, in quadrupedal monkeys, as shown by the figures 9, 40, 41 and 42. In lemurs and *Megaladapis*, for example, the nuchal surface stands nearly perpendicular, immediately above the occipital joint (figures 6 and 39), but in the habitually upright tarsier this area for the attachment of the neck muscles occupies about the same direction as it does in recent man (figure 2). An occipital crest, like a sagittal one, develops only with and after the completion of general growth in fully adult monkeys and apes. These crests attain their greatest height in old male gorillas, in which they coalesce in a backward projection at the top of the occiput, thus adding much to the height and length of the head. In many chimpanzees and gibbons, on the other hand, these crests can be totally lacking, even in old males.

The size and shape of the brain case are primarily determined by the evolutionary status of the brain and the size relations of its parts. With the trend to enlarge the relative size of the brain the neurocranium tends to become in general more globular, as the form which contains the optimum volume in the most economic bony enclosure. This explains also the general trend to increase the proportionate height of the brain case from its low forms in lemurs to the much higher form in hominoids (figures 39 and 40). The high forehead of modern man is also a direct result of the unique development

97

of the human frontal lobes. Individually, however, the brain does not *have* to determine the shape of the skull very closely, as is evident from the ease with which the head can be deformed artificially during infancy, as in the American 'Flathead' Indians or in the many other tribes which practised occipital flattening. It is furthermore shown by the occurrence of abnormally early closure of the sagittal suture, resulting in extremely narrow skulls (= scaphocephaly), or the early obliteration of the coronal sutures, leading as compensation for the reduced growth in length to abnormally high skulls (=oxicephaly), and this not only in man, but also occasionally in wild monkeys and apes without the least effect on the function of the misshaped brain.

The outer form of the neurocranium can often obscure the size and shape of the brain cavity through the formation of the crests, discussed above, and on account of extensive sinus formations in the cranial walls. The latter develop not only in the maxillary and ethmoid bones, but also quite frequently in the supraorbital region and the skull base, and even the temporal bones can contain elaborate pneumatic spaces to economise in bony material while attaining a mechanically efficient construction. One of these pneumatic parts is the mastoid process, formerly and erroneously assumed to be an exclusively human specialisation. These processes behind the ears serve for chiefly the attachment of the sterno-cleido-mastoid muscles which effect certain movements of the head. In recent man these processes appear early in infancy and soon reach their final and very considerable size. In the great apes, especially the African ones, identical processes develop much later in life, but can ultimately become fully as large as in man. Other primates lack these processes, so that the corresponding muscles are simply attached to the unspecialised mastoid wall.

With the progressive migration of the face underneath the braincase in simian primates the lower jaw had to increase the height of its 'ascending' rami to reach the mandibular joints at the cranial base, far above the plane of occlusion of the upper and lower teeth (e.g. figure 42). In all prosimians with protruding faces the fulcrum of the lower jaw still lies nearly in the general direction of dental occlusion (e.g. figure 1), so that their jaws operate much like the blades of simple scissors, rather than on the principle of some types of nutcrackers. The elevation of the mandibular condyles above the dentition in the higher primates develops only during postnatal life.

Of the innumerable further specialisations of the various cranial bones only those of the premaxilla can be briefly discussed here, since they will be mentioned also in the chapter on growth and development. These paired premaxillary bones, which carry the upper incisor teeth, are comparatively small in all recent prosimians with their small incisors, except in the aye-aye in which they have become very large in connection with their uniquely specialised, large front teeth (figures 1 and 7). Among monkeys the premaxilla

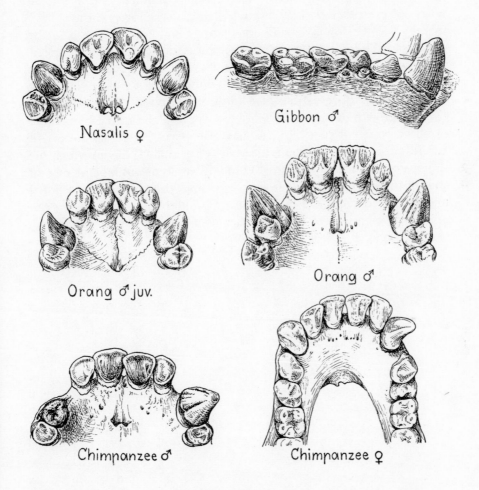

Figure 43   Examples of crowded teeth in some wild catarrhines (after Schultz, 1966).

is better developed, extending at least to the nasal bones and in many species even alongside the latter all the way to the frontal bone. Among the hominoids, with their comparatively large upper incisors, the premaxillaries have become correspondingly wide and often also form a relatively high part of the face underneath the nasal aperture. In addition, the premaxillaries of many primates carry the upper incisors forward and thus leave gaps in the dental row, known as diastemata, into which fit the lower canines when the mouth is closed. These diastemata are generally best developed in males with their large canine teeth, but occasionally they can be nearly or even entirely lacking, in which case the lower canines have merely become somewhat divergent to permit full occlusion in the dentition. These diastemata, furthermore, do occur even in specimens with congenitally absent or lost lower canines and they have been found in not a few human skulls without, of course, protruding canines. These observations give additional support to the conclusion that the growth of the jaws is not very closely determined by the size of the completed dentition, as has also been shown by means of careful measurements on a large series of primate skulls in which the dimensions of the jaws were found to vary quite independently of those of the tooth rows. It is not surprising, therefore, that crowded and impacted teeth are by no means limited to civilised man, but also fairly common in many species of wild primates, as shown by the examples in figure 43.

Chapter 7

Oral Cavity

Teeth

THE TEETH of lower vertebrates are generally quite uniform in shape in contrast to the conditions in mammals whose teeth have as a rule become·differentiated into regionally distinguishable types to serve special functions. That the few exceptions, such as in recent toothed whales, have resulted from secondary reductions, can be demonstrated by fossil forms. In typical, unspecialised dentitions of placental mammals the comparatively simple front teeth, or incisors, serve chiefly for cutting. They are usually followed by projecting canine teeth, used for grasping and piercing, and behind these by a row of more complicated teeth for the preparation of food by chewing, the so-called premolars and molars. Since teeth consist of more durable substance than bones, they are most frequently preserved in a fossilised form and have thereby provided us with the most numerous, and sometimes the only, documents for recognising the many widely diverging trends with which mammals have acquired different modes of life and have adapted their dentitions accordingly. Here it can merely be mentioned that the dentitions of primates have remained more conservative than those of many other mammalian orders, among which some far more extreme alterations of the original condition have evolved. It is generally assumed, for sound reasons, that the earliest placental mammals, including the ancestral stock of primates, possessed in each half of each jaw three incisors, one canine, four premolars and three molars, thus having had a primitive dental formula of $\frac{3\ 1\ 4\ 3}{3\ 1\ 4\ 3}$.

This full complement of forty-four permanent teeth soon underwent at least some degree of numerical reduction in all primates, except in the ancient treeshrew-like *Anagale* of disputed primate affiliation and, possibly, in some Eocene tarsioids. Following the old definition of Linnaeus the possession of two incisors in each half of each jaw had long been cited as one of the main diagnostic characters of the primate order, but gradually many exceptions have

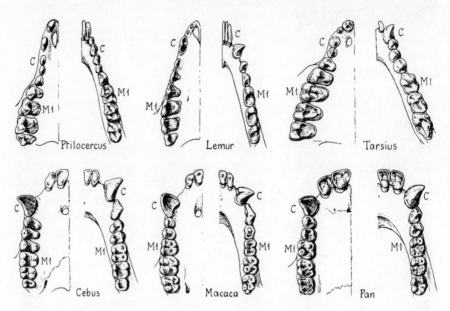

Figure 44   Right upper and lower permanent dentition of some prosimian (above) and simian (below) primates, C = canine, MI = first molar.

become known. For instance, the extinct as well as recent tree-shrews all still have three lower incisors on each side (figure 44). The general trend among primates to reduce the original dental formula to only two incisors, or even fewer, manifests itself clearly and has repeatedly reached extreme forms. In the majority of lemurs the upper incisors have dwindled to mere rudiments (figure 44) and in some species one or both of these teeth can vanish altogether as seen in many individuals. The recent *Lepilemur* and the subfossil *Megaladapis* show the complete loss of upper incisors and tarsiers have normally only one lower incisor left (figure 44). In the most highly specialised dentition of all primates, that of the aye-aye (figure 7), the incisors of both jaws have been changed into single, huge gnawing teeth, resembling those of rodents also by being only partly covered with enamel and probably being of continuous growth. Similarly specialised incisors had also developed in some not very closely related fossil prosimians. The adaptability of teeth to special functions is nowhere better shown than in the lower incisors of lemurs and some other recent prosimians which have become very slender and nearly horizontal, and close to which the lower canines have not only moved, but also acquired very similar shape, so that the total of these six lower front teeth form a very serviceable projecting comb for the care of the fur, or as a scraper, useful in eating (figures 1 and 44).

All these specialisations in the number, shape and direction of the incisors refer to prosimians. Compared with this remarkable variability, the incisors of all simian primates are far more uniform, all normally having the same number and usually a spatulate form with a fairly straight cutting edge. The upper middle incisors of the great apes have the minor distinction of being relatively broad. In recent man there occurs a familial tendency to reduce and even lose the upper lateral incisors in small percentages of cases.

Of the canines, both the upper and the lower ones have become lost in *Daubentonia* and only the lower ones in *Indri* as well as in some few fossil prosimians. That the lower canines of still other prosimians can acquire the size and shape of incisors has already been mentioned. In the latter case the most anterior of the premolars has sometimes become enlarged to take over the function of a canine. In the majority of primate species of all major groups the canines reach well above the other teeth as long and strong instruments for holding and cutting food as well as effective weapons. This wide-spread evolutionary trend is especially marked in the upper canines and usually much more pronounced in males than in females. In fully grown males of most monkeys and, particularly, of baboons the upper canines have become greatly hypertrophied to form long daggers (figure 73). Males of the recent great apes also possess very powerful upper canines, surpassing in their dimensions those of females in *all* gorillas and orang-utans, whereas among chimpanzees only in the majority of the cases. The gibbons and siamangs are distinguished by canines of practically the same great length in both sexes and some species of marmosets also show no noteworthy sex difference in the (moderate) size of their canines. The same can apparently be said in regard to the fossil australopithecids and, of course, to recent man, who also has small canines in both sexes. That the latter condition represents possibly a secondary reduction is indicated by the seemingly unnecessary length of the roots of these teeth. The long canines, present in the majority of

primates, limit lateral movements of the mandible during mastication, though the intensive wear of these teeth soon permits
considerable play. The posterior edge of the crown of a long upper
canine acts with the anterior part of the first lower premolars like
blades of shears.

Of the original four premolars the most anterior one had already
become eliminated very early in primate evolution. In the Indriinae
and again in all catarrhine monkeys, all apes and man even, two of
these teeth have disappeared, leaving only two premolars in each
side of each jaw, corresponding usually to P3 and P4 of the ancestral
formula. By far the most extremely reduced dentition, that of the
aye-aye, contains only one small premolar in the upper jaw and
none in the lower one (figure 7). Among numerous species of lower
primates the last premolars tend to become enlarged and more
complicated in form, thus resembling molar teeth.

The number of molars has remained constant in all primates,
except for the great majority of the Callithricidae, which have lost
their third molars in both jaws, *Callimico* being the only member of
this family still with three molars. Intraspecifically one can recognise
a clear trend toward the reduction or even loss of the last molars in
many cebids as well as in recent man. Among the former this is most
pronounced in some local populations of spider monkeys, which
have been found to lack last molars in up to fifteen per cent of the
cases, while additional specimens have at least one to all four of
these teeth greatly reduced in size and number of cusps and roots.
Series of *Cebus* and of *Saimiri* skulls often show similar conditions
also with significant percentages of occurrence. That the last molars
of man are quite commonly much reduced in size, delayed in
eruption, or congenitally missing, is a clear indication that they
could become gradually eliminated in the process of shortening the
human dentition. In the large-jawed great apes, on the other hand,
the third molars usually equal or surpass the others in size and even
supernumerary molars develop much more often than in other
primates. It must be pointed out, however, that cases of additional
molars, or more rarely of canines, naturally cannot be regarded as
atavisms, since they do not repeat an ancestral condition. To be
consistent, the sporadic appearance of supernumerary premolars or
incisors should also not be interpreted as individual retentions of
these formerly more numerous teeth. That all such cases have
resulted from localised aberrations in the prenatal development of

the dentition is a far more plausible explanation in view of the many instances of incomplete duplication of single teeth. All such occurrences are comparable to the many other examples of localised twinning, such as the not rare supernumerary digits, which vary from mere small buds, branching from one phalanx, to complete extra fingers or toes. Individual reductions in the size and number of teeth, which are generally more common in primates than are increases, can be of evolutionary significance. This is evident not only from the various reductions of the original dental formula, acquired by nearly all primates, but should also be admitted in regard to the hereditary variations in at least the size of molars and of incisors, including their occasionally complete lack. It is by way of such individual occurences that the general reductions have been attained which characterise entire taxons. In connection with the relations between ontogenetic and phylogenetic processes it is also of interest that the deciduous first dentition, which develops very early, shows fewer individual variations as well as generally less marked specialisations than the later permanent dentition.

It would lead too far to describe here also the many and intensively studied technical details of the finer construction of primate teeth, which have been most helpful in the interpretation of fossils and their interrelationships. The above notes and accompanying illustrations must suffice as support of the general statement that incisors, canines and even premolars have tended to deviate more from their basic patterns than have the molar teeth, which have usually retained their fundamental primate characteristics more tenaciously – probably on account of their more uniform functions.

Tongue

No other muscular structure can change its position and shape as much and as readily as the tongue which serves a variety of functions in all primates. Being highly movable the tongue rolls food around for mixing with saliva, shifts it to the teeth for grinding, renders effective aid in removing food remnants from the cheeks and teeth, and with its changing positions it can modulate sound. As a thoroughly innervated muscle complex the tongue acts also as a delicate instrument of touch. This ability is of particular advantage to the many nocturnal prosimians which regularly explore objects in front of the mouth with the extended tongue. The epithelial lining

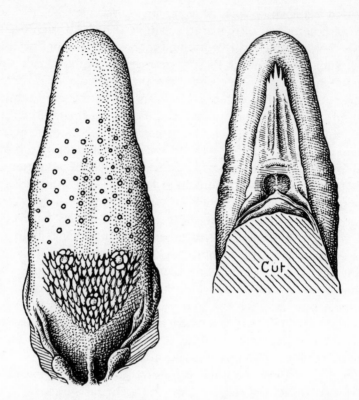

Figure 45   Upper and lower aspects of the tongue of a lemur, showing the many small fungiform and few large vallate papillae, the latter hidden among the conical thorny papillae on the pharyngeal surface. The denticulated sublingua with its median ridge shows clearly on the lower aspect.

of the dorsal side of the tongue contains numerous small and fewer large papillae which act as receptors of the vital taste sensation. The large so-called vallate papillae, which are mostly toward the back of the tongue and vary in their never large number, are supplied with the greatest concentration of taste buds. In many prosimians they are surrounded by tough conical papillae, directed toward the rear and thus holding food about to be swallowed (figure 45).

Of special interest is the so-called sublingua, a flat fleshy plate attached to the lower surface of the tongue, except for its tip and lateral margins (figure 45). This structure is found in marsupials and certain other lower mammals and exists in all prosimians and, as mere remnants, in the more primitive types of platyrrhines. In

Lemuridae, Lorisidae and Galagidae this sublingua is remarkably well developed, reaching far forward, stiffened by a midsagittal ridge, and bearing at its apex a varying number of slender serrated extensions, used in cleaning the interstices of the comb-like lower incisors and canines. Some of the details of this specialisation for a

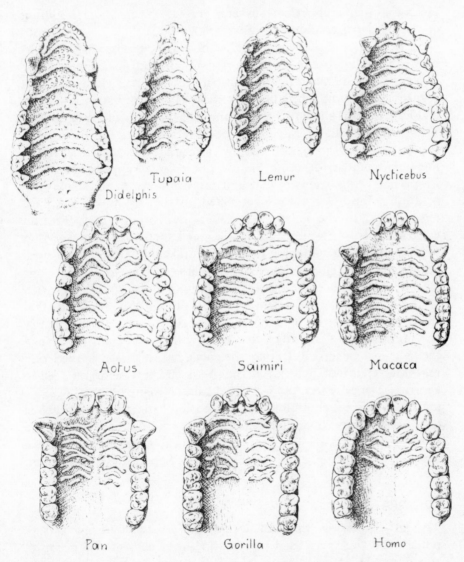

Figure 46   The palatine ridges of a marsupial (American opossum) and of various primates.

highly limited purpose have not developed in treeshrews, the aye-aye and the tarsiers which lack also the peculiar arrangement of the lower front teeth, typical of the other prosimians.

Palatine Ridges

Nearly all mammals bear transverse ridges, or rugae, on the lining of the hard palate. Some ungulates have twenty-six or sometimes even more thick rugae, covering the entire palate, whereas in whales they are totally absent. In many marsupials and insectivores these rugae, though high and tough, number only seven or eight and form regular connecting ridges between most of the corresponding right and left teeth, as shown by the example of the opossum in figure 46. This arrangement of ridges has been retained with little modification by the great majority of lower primates, of which six types are illustrated in the same figure. These rugae appear very early in prenatal life, remain constant in number, main shape and continuity, and are always restricted to the region of the hard palate only. In the mostly edentulous newly born they form an essential aid in grasping the nipple and later in life they serve in holding and even crunching food in the oral cavity if they are as well developed as in all prosimians. In a variety of monkeys the ridge pattern tends to become somewhat irregular and to no longer reach all the way back to the last molars. It is among the hominoids that the latter trend has become most pronounced as far as the irregularity and asymmetry of the ridges and their withdrawal from the rear of the palate is concerned. In man the rugae are restricted to the front, reaching often merely as far as the premolars and being reduced to an average of only four on each side. Like so many other degenerating structures which no longer serve a useful function, the rugae of the higher primates individually vary far more than they do in monkeys and lemurs. In man, for instance, their numbers range between two and eight and in chimpanzees between five and fifteen.

Pouches

To this chapter on the oral cavity may be added a few further remarks on the specialised pouches in the cheeks of one group of Old World monkeys. Between the dentition and the lips and cheeks lies a vestibule of the oral cavity which can be widely expanded when filled with air or fluid with the mouth closed. All species of the subfamily of Cercopithecinae have acquired cheek pouches as specialised extensions of the vestibule, as shown by the example in figure 11. These pockets serve as convenient storage places for food which is not being masticated immediately, but will be removed later through the contraction of the overlying musculature (figure 77). In older macaques and baboons these pouches, when jammed full, can attain an amazing size, reaching far below the lower jaw and sometimes requiring the aid of fingers to be emptied. It is noteworthy also that these pouches are already developed during prenatal life, indicating that this specialisation is well established in the cercopithecines, though lacking in all other primates at any stage of growth.

Chapter 8

Skin and Hair

MANY aspects of the skin and hair are referred to most appropriately in the chapters on age changes, special senses and secondary sex differences so that here it remains only to deal chiefly with other kinds of specialisations of the skin and its derivatives in primates. It is typical of the generally conservative nature of this order that none of its members has developed any of the extreme dermal changes found in a great variety of other mammals, such as the armour of armadillos and of an extinct giant ground sloth, the scales of pangolins, the quills of echidnas, hedgehogs and porcupines, the complete hairlessness of whales, or the many different types of horns and antlers of ungulates. The chief and well marked specialisations of the skin, found exclusively among primates, are the periodic changes of the female sex skin in many different catarrhines and the widely distributed occurrences of very varied and extremely colourful patterns of pigmentation of skin and hair by which some species are rendered very conspicuous and certainly not camouflaged.

No primates had to acquire thick, heat-conserving layers of subcutaneous fat, typical of many arctic mammals. The few monkeys surviving in cold winters manage with slightly longer hair, besides close mutual contact during rest, and Eskimos manage with furs and fire. The normal and striking accumulation of fat in the proximal half of the tail of *Cheirogaleus* disappears gradually during the dry season, when these prosimians remain in a torpid state, and is therefore merely a periodic and localised phenomenon, useful in providing water through chemical changes. Such localised fat storage occurs also in many other mammals, as in the humps of camels and zebus and, particularly, in the tails of some species of marsupials, insectivores and rodents, as well as in some breeds of domestic sheep. The regionally restricted fat deposit in the buttocks of adult females of such human races as Hottentots, Bushmen and Andaman islanders, known as steatopygism, is due to some endocrine factor and seems to have no special adaptive value. Fully grown, healthy male baboons usually accumulate thick pads of subcutaneous fat exclusively in the region surrounding their ischial callosities for unknown reasons. The fatty cheek pads which develop after sexual maturity in chiefly male orang-utans and the regular crown pads of fat and tough connective tissue fibres on top of the head in old gorillas have also nothing to do with heat regulation. The cheek pads serve probably as 'impression ornaments', similar to the conspicuous paranasal corrugations of mandrills or the heavy pads on both sides of the face of drills, and the crown pads as cushions between the scalp and the hard edges of the occipital and sagittal crests at their meeting behind the cranial vertex.

Among the many specialisations of the skin in primates the unique formations of a bare area on the chest of geladas are particularly noteworthy. Both sexes of this monkey, popularly often referred to as 'bleeding heart baboon', have two hairless triangular zones on the throat and chest which are generally of a more or less bright scarlet colour. In older females these conspicuous patches become periodically surrounded by rows of cutaneous vesicles, or partly pendulous warts, more romantically giving the appearance of pearl necklaces (figure 47). In sexually mature females the changing size and colour of these structures and the accompanying fluctuations in the colouration of the enclosed areas are known to be correlated with the menstrual cycles, so that this remarkable specialisation is equivalent to the 'sex skin' of many other catarrhines, except for its location, appendages and the fact that it is not fully limited to females. The occurrence of permanent, localised, bare and red skin in males is found not only on the chest of geladas, but also on the face of several species of macaques and on the face as well as forehead of uakaris (Cacajao) (plates 33, 39). In mandrills the central part of the nose from root to tip is brilliantly red and flanked by conspicuously contrasting blue zones on both sides, this unique colour design of the skin being most marked in males. Here it must also be mentioned that in a variety of monkeys, as for example geladas, the upper eyelids remain unpigmented and, when fully exposed by raising the brows above, give the animals a startled expression, evidently used for intimidating antagonists. In other species of Old World monkeys, especially the 'spectacled langurs', the skin of the entire orbital region is chalky white in the midst of an otherwise dark face and thereby accentuates the staring appearance

Figure 47 Adult female gelada baboon showing the chains of vesicles reaching from the two round caruncles at each side of the neck to below the breasts, which are separated by the perpendicular fold flanked by the bulbous nipples (after a photograph by Matthews, 1956).

of their shiny-dark eyes. By far the majority of the extraordinarily colourful and sharp-cut designs, characterising so many primate species, depend not on the skin, but on the colour of the hair in its often surprisingly great regional differences, as will be briefly described later on.

Mammalian skin is generally quite elastic or, rather, stretchable. This is very evident from the sometimes huge pendulous breasts of multiparous and long-lactating women or the grotesquely enlarged

Figure 48   Head of a sub-adult male chimpanzee (after a photograph).

ear lobes of some human tribes. Skin, however, does not always shrink in area with a reduction of the contents covered by it, but instead forms wrinkles and folds. Such folds are most constant at joints where skin becomes alternately stretched and bent. These are the skin folds on our knuckles, the inguinal folds between the thighs and abdomen, the flexure folds of the palms and the individually variable folds and wrinkles of our eyelids. Nonhuman primates possess the same types of main flexure folds as we have, though naturally they vary in their precise location, direction and depth according to the species differences in the kind and extent of the movements of their joints or with the amounts of alternating change in a given area of skin. The skin over such structures as the voluminous throat pouches of orang-utans and siamangs shows wrinkling and deep folding particularly well after deflation. The amount of wrinkle formation in the face contrasts very strikingly in most monkeys and all apes with the condition prevailing in man before senility by being not only more marked in the former, but also by developing much earlier, so that even newly born and infantile monkeys and apes resemble in their faces emaciated and toothless old men (figures 48 and 58). To some extent this may be due to the fact that nonhuman primates under natural conditions lack the padding of subcutaneous fat which appears in man before birth and normally persists at least through infantile life. It is of interest also that the skin of wild monkeys has been found to be endowed with a remarkable capacity for the repair of injuries. Even extensive and gaping cuts and tears usually heal by themselves in a surprisingly short time and without leaving noticeable scars. This is most clearly seen on the trunk and limbs, where the skin is more loosely knit to underlying structures than it is in man. Wounds on the outer ears, whose skin is closely attached to the deeper tissues, heal poorly and usually remain as lasting tears. Even umbilical scars, so conspicuous in all human beings throughout life, tend to disappear completely during juvenile growth in a great many nonhuman primates.

The countless minute glands in the skin of all primates, which play important roles in heat regulation, secretion, etc., have as yet been investigated in only a few species. Lacking systematic information on the distribution, density and other aspects of these glands, it can merely be stated that apparently most primates can and do sweat over at least many parts of their body surface. The most important of the highly specialised skin glands – the milk-secreting mammary glands – will be discussed in the chapter on reproduction, where they are of greatest interest. As in a wide variety of other mammals, some primates have also developed other specialised skin glands which usually secrete some odoriferous substance for marking pathways and territories, or simply for the recognition of fellow members of a species. It seems to be most likely that some of these glands also have a sexual significance because they are often much better developed in one sex, or even limited to it. Such glands are naturally most frequent among the prosimians, which still have a fairly well functioning sense of smell, but can be located in very different parts of the skin. In various prosimians and some marmosets they are found in the anal or genital region, in treeshrews, woolly monkeys, spider monkeys and most male orang-utans in the middle of the chest, in male sifakas under the chin and in both sexes of avahis on the sides of the neck. In some species of lemurs there occur large specialised glands on the upper arms as well as on the forearms associated with curious horny spurs, which are more strongly marked in males than in females.

The palms and soles of mammals, whose distal limb segments have not become extensively altered, bear more or less distinct pads chiefly at the tips of the digits and between the digital bases as well as proximally on both sides. These 'walking pads' serve to protect the deeper soft structures and act as shock absorbers where the skin would be pressed against skeletal parts. They appear very early in prenatal life, as shown by the examples in plate 6. In many different mammals, including most lower primates, the pads remain as quite sharply defined, thick cushions throughout life, but in higher primates they soon become indistinct, though their primary existence and influence can still be recognised in the arrangement of the main flexure folds, which naturally originated between, rather than across, the pads. The earliest stages of the delicate normal development of the primary and possible accessory pads can apparently become easily disturbed by certain chemicals or by virus infections, to which the mother had been exposed during the first part of her pregnancy. As a result the pattern of the flexure folds will also become altered, producing for example in man the rare and misnamed 'simian palmar crease', usually seen in 'Mongolian idiots', which represents a teratological but not an atavistic condition.

In the skin over the pads there can appear more or less numerous, minute and parallel ridges, which in the primates become very extensive and form definite, complicated patterns, known as dermatoglyphics, which in man are widely used for individual identification. This highly specialised ridge-bearing skin is completely free of hair and of sebaceous glands, but contains a great many hypertrophied sweat glands and is well supplied with sensory nerves. In this way it forms an ingenious device not only for tactile

Figure 49   Hand of an adult baboon showing the dermatoglyphics and pads in a very similar arrangement as in the hand of the langur in figure 25 in spite of the very different proportions of this plump hand.

sensations, but also for improved adhesion to supporting surfaces, much like the corrugations on automobile tyres. Cutaneous ridges on palmar and plantar pads have appeared to a limited extent and at best in a rather coarse form among some marsupials, insectivores, edentates, rodents and carnivores, having most likely developed independently and are never present in all species. The order of primates is distinguished by being the only one in which all living species possess these ridge formations, though with widely differing degrees of perfection and completeness. In the great majority of prosimians the specialised ridged skin is still limited chiefly to the pads, while between the latter there are mere wart-like elevations and in some species the sole under the heel has not even become denuded of hair. The palms and soles of nearly all monkeys are completely covered with dermatoglyphics which contrast with those of prosimians also by being usually more delicate and much less diversified in their general arrangement among the various sub-groups. This is very evident from a comparison of the slender hand of the arboreal langur in figure 25 with the plump hand of the terrestrial baboon in figure 49, both of which have the same general ridge pattern in spite of their strikingly different proportions. Among all the recent hominoids the ridged skin has become most highly developed, but differs very markedly in gibbons, great apes and man, thus showing clearly the divergent evolution of these groups and not different stages of perfection. Individual variations in the form and relative frequency of all the detailed ridge patterns of loops, whorls and spirals are fully as common and marked in apes as in man and change during growth as little in the former as in the

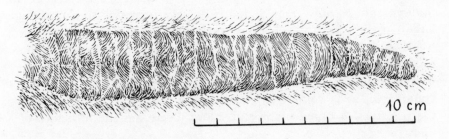

10 cm

Figure 50   The specialised tactile skin on the ventral side at the distal end of the tail of an adult *Lagothrix*.

117

latter. It is highly interesting that the skin on the distal half and ventral side of the prehensile tails of howler-, woolly- and spider-monkeys has also developed regular dermatoglyphics for better friction as well as greater sensitivity of these specialised 'fifth hands' (figure 50).

Claws are ancient specialised appendages of the skin at the tips of the digits. They are a characteristic reptilian feature and have persisted in many mammals as useful instruments in climbing, digging and scratching as well as fighting. The typical claw consists of two layers of horny substance, of which the deep and thick one is particularly tough and durable while the thin, superficial layer serves as a protective, smooth cover. Most claws are strongly curved from their bases to the sharp ends, compressed from side to side, and quite closely moulded to the terminal phalanges, which assume a corresponding shape. It is due to the latter fact that even fossil remains often permit conclusions regarding the former existence or lack of claws. Among primates claws have been retained on all digits by the recent treeshrews and on some of the digits by the aye-aye as well as some other prosimians and all Callithricidae. These claws still have a deep horny layer, more or less reduced in extent and thickness, projecting with their curved, sharp points well beyond the fleshy ends of the digits and serving the small animals in clinging to the bark of trees. In all other recent primates claws have become replaced by nails, which are regarded as degenerate claws, having lost most or all of the deep horny layer and sharp points. Nails also do not fit as closely over the terminal phalanges, even though they are quite strongly curved transversely in most monkeys. Their primary function is mechanical support opposite the digital touch pads and, of course, they are indispensable for scratching and picking. In tarsiers the nails have become reduced to mere vestiges, undoubtedly because their digital pads have increased to singularly large flat discs with which these dwarfs among primates gain firm holds on even perpendicular smooth surfaces. With the degeneration of single digits in so many different primates, discussed before, the corresponding nails also dwindle to small remnants or disappear completely, as is commonly seen on the thumbs of guerezas and the first toes of orang-utans and, more rarely, on the fifth toes of man.

It is commonly taken for granted that the striking lack of hair on most of man's body represents a unique specialisation among primates. This extreme condition, however, appears much less impressive, if one considers the enormous generic differences in the density of hair, according to which many nonhuman primates have also undergone very marked reductions in their hairy covering in contrast to the surprisingly dense coats of hair in other species, living under very similar, if not identical, environmental conditions. This is best shown by the data in table 3. For instance, a square

Table 3   Average numbers of hairs on a square centimetre of skin at vertex, middle of back and middle of chest in adult simian primates according to determinations by the writer and for *Callithrix* by Wettstein (1963).

Genus	Scalp	Back	Chest	Genus	Scalp	Back	Chest
Callithrix	4010	2380	1870	Cercopithecus	1880	1670	240
Aotus	3590	2880	1210	Erythrocebus	710	1550	130
Alouatta	1190	1100	220	Presbytis	1060	710	160
Saimiri	2010	1650	600	Nasalis	670	1070	380
Cebus	1230	1130	440				
Lagothrix	2550	2420	720				
Ateles	960	650	190	Hylobates	2100	1720	600
				Symphalangus	710	430	260
Macaca	650	550	70	Pongo	160	170	100
Papio	640	655	135	Pan	180	100	70
Cercocebus	850	980	260	Gorilla	410	140	5
				Homo	300	0	1

centimetre of skin from the middle of the back supports less than two hundred hairs in the great apes, but well over two thousand in various platyrrhine monkeys. The corresponding hair densities at the centre of the scalp average less than two hundred in orang-utans and chimpanzees, about three hundred in adult human beings (often more in Europeans and much less in Mongolians), but over two thousand in gibbons and many New World monkeys, with the highest density of four thousand in a marmoset. Even on the chest, where the hair density is invariably most reduced, the relevant data in the table differ far less between man, the great apes and some catarrhine monkeys on one side and gibbons and some platyrrhines on the other side. We have as yet no plausible explanation for these surprisingly wide discrepancies in the density of the hair of wild simian primates. Some macaques, langurs and gibbons are of very similar body size, live in the same jungle off largely the same diet, yet gibbons have clearly the densest coat of hair. Spider monkeys and woolly monkeys are also of practically the same size and adapted to the same tropical climate, but the latter show a much greater hair density than the former without any apparent disadvantage. Siamangs grow not nearly as many hairs on a given area of the body as do gibbons, which live in the same regions. In view of these very striking genetic differences between even closely allied species or genera the various attempts to account for the extreme reduction of hair in man as a symptom of domestication or as a result of sexual selection appear more fanciful than convincing. The many kinds of regionally limited loss of hair among primates are equally futile to explain as selective advantages. The practically bare foreheads of uakaris as well as of some orang-utans and chimpanzees, or the nearly bare chests of gorillas and geladas have evolved alongside densely covered foreheads and chests in closely related species. Toque macaques have conspicuous bald spots at the vertex, from which dense hair radiates to the brows, ears and neck. In many baboons a wide zone surrounding the ischial callosities bears almost no hair. Localised increases in hair length are frequent species characters among primates and are often sex-limited. One need only mention the thick hairy mantles of male hamadryas baboons and geladas, the flowing long hair on the back and tail of some species of guerezas, and the grotesque halo of hair surrounding the face in some macaques, especially *M. silenus*, and langurs, the long bushy hair near the ears of certain marmosets, the very long hair hanging from the scrotum of siamangs, the great variety of imposing beards and moustaches of some species of marmosets and of guenons (plates 18, 19, 24, 34, 35) and, of course, the hairy facial decorations of man, generally best developed in the white race.

As a rule the hair of primates tends to point in a cephalo-caudal direction from the brows across the vertex all the way to the tip of the tail and on the limbs more or less distally, as shown by the direction of the foetal hair of a macaque in figure 51. This streamlined arrangement offers the least resistance to forward locomotion and hence exists in its clearest form in such aquatic mammals as otters and seals. Among primates many localised changes of this general hair direction have developed for unexplained reasons. For instance, the hair on the throat often points upward and that on the cheeks commonly forms whorls or meets from opposing directions to rise in lateral crests which, if the hair is long, become very con-

Figure 51   Foetus of a rhesus monkey with diagrammatic representation of the approximate direction of the main hair tracts of lanugo, indicated by the lines and arrows.

spicuous in some species. Further specialisations in hair direction, producing patterns of convergence or of radiation, are most frequently found on the upper parts of the chest, in or near the axillae, on the medial sides of the forearms and anywhere on the

121

scalp from the brows to the occiput, unilaterally or bilaterally. All these and many more local deviations from the general main type of hair direction appear already in the earliest stages of hair development and can represent merely individual variations or such prevailing ones that they have become regular racial or species characters. Nature's versatility in this respect had been thoroughly demonstrated in a huge monograph by a famous German anatomist, G. Schwalbe, who had described so many detailed variations of hair tracts in a large series of unborn monkeys and apes that it became hopeless to summarise them.

It is readily seen from the endless possibilities for combining variations in the density, length and direction of hair with those in hair colour that there is no difficulty in marking species, races, sexes and often even ages very differently solely by means of their coats of hair. As is mentioned also in connection with colour vision, the range of pigments in the hair of primates surpasses that of all other mammals, since it extends from pure white to real black, via many sorts of red, blue, yellow, brown and even what appears as green through alternating yellow and black or brown rings on hair shafts. While numerous primates are more or less uniformly coloured all over in mostly sombre tones, a great variety of other species bear strikingly contrasting colour patterns which render them very ornamental and highly conspicuous. Some species of tamarins, of guenons and of langurs have white upper lips in otherwise dark faces, the 'putty-nosed guenon' has shiny white hair on the tip of the nose, the white-handed gibbon wears not only white gloves, but also a white band across the eyebrows. The skins of some guerezas and those of the 'golden monkey' (*Rhinopithecus*) are of such decorative beauty that they have nearly led to the extinction of the species through intensive hunting in parts of their distribution. Some of the rarer species of tamarins and marmosets are brilliantly orange-coloured over parts of their bodies, while others are almost white and still others bear alternating patches of different colours on their backs and tails. The wealth of different colour designs among the platyrrhines of the Amazon region has been most impressively shown in forty-two beautiful coloured plates by Eladio da Cruz Lima, who was a devoted naturalist and former judge of the Supreme Court of the State of Para in Brazil. Various examples of gaily coloured primates are shown in the plates 32 to 39. Many marked individual variations in the pigmentation of hair have been recorded in the literature and even proposed as new subspecies. For instance, among the normally black-haired chimpanzees specimens with reddish hair have been found repeatedly and once a chimpanzee born to normally coloured parents had pale straw-coloured hair, turning later to light brown. Albinos are known to occur sporadically not only in man, but in all major groups of primates, as they do among other mammals from opossums and mice to giraffes, elephants and whales.

Far more interesting to modern biologists than the bare description of the multitude of different patterns of hair colour would be a consideration of their possible cause of origin and effect on survival. If some species of primates thrive in uniformly dark coats of hair, which conceal them as mere silhouettes in the gloomy jungle light, why should so many other species wear blatantly advertising colour designs, though differing little in habits and habitats? Much more field research is required for solving this puzzling problem.

Chapter 9

Brain

THE INVESTIGATION of the evolutionary specialisations of the brain has been much helped by the fact that the brain, unlike other soft parts of the body, is enclosed in a bony capsule which, when persisting in a fossilised state, preserves reliable indications of the general size of the brain and often even of the proportionate sizes of some of its parts and of their spatial relations to each other. In many mammals, including most lower primates, the inner surface of the neurocranium shows the configuration of the brain surface in sufficiently close detail that much can be learned of its evolutionary stage. In pongids and hominids, however, casts of the brain cavity of the skull do not reveal the highly convoluted brain surface with all the desired minuteness on account of the membranes, blood vessels and fluid which are interposed between the skull and brain and together amount to roughly ten per cent of the cranial capacity. Even with such limitations the new science of palaeoneurology has been able to add much valuable, direct evidence in support of conclusions, derived from comparative studies on the brains of recent species, regarding the general trends of evolutionary specialisation in the primate brain. To avoid a lengthy discussion of highly technical details, this brief chapter will deal only with some of the topics appertaining to the general and chiefly macroscopic conditions of the brain.

The order of primates is distinguished among mammals by its tendency to develop brains which are extraordinarily large in relation to the total size of the body and which furthermore show a gradual expansion and elaboration of the cerebral cortex. The latter serves primarily for the reception of sensory impulses and their transformation into actions forming suitable behaviour patterns. The progressive differentiation of the cortex in primates has increased their ability to perceive as well as to react to environmental stimuli with improved versatility in connection with the strict distribution of the many specialised functions among different cortical areas. The disposition and proportionate extent of these quite sharply differentiated areas represent a complex mosaic which in many respects is characteristic of primates and becomes most pronounced in the higher members of the order.

Corresponding to the different roles of the senses in the life of primates, the neural apparatus of vision is especially well developed and with it the equally important tactile sense so essential for these animals, primarily adapted to an arboreal existence. On the other hand, the olfactory cortex has generally undergone a comparative reduction, which has progressed furthest in all the higher species which benefit little from their sense of smell. The auditory area of the cortex is moderately well represented without any outstanding tendency to increase or decrease in accordance with the conservative reliance on hearing in all primates. The so-called association areas, which are concerned with the interrelations between the sensory and motor projection areas, become highly developed in the cerebral cortex, particularly in monkeys, apes and, most of all, in man. In connection with all these complex functions and their increasingly close localisation in definite cortical areas the surface of the brain tends to become folded into more and more elaborate systems of convolutions and intervening sulci. These folds, naturally, increase the specially important surface area without adding to the total brain volume. They are still lacking, barely indicated or only shallow in small, primitive species, but become more numerous and deep in the larger ones and reach their extreme development in the higher primates during postnatal life, as shown by the examples in plate 9.

Hand in hand with the expansion of the cortex some of the main lobes of the primate brain have undergone specially intensive development. For instance, the occipital lobe becomes progressively enlarged and extended toward the rear to accommodate the visual cortex and thereby shifts more and more over the parts of the brain underneath. The temporal lobes of primates are also distinguished by a clear trend of increase in size and convolutions, though not due to the auditory cortex, which occupies only a small part of them, but possibly for the storage of visual memories and the reception and integration of many sorts of impulses. The frontal lobes with their highly complex and as yet not fully explored functions share in the remarkable development of the brain among primates by their progressive growth and differentiation from monkeys to apes and recent man, resulting finally in our high foreheads. The olfactory bulb, as the receiving centre for nasal impulses, forms a conspicuous and projecting part of the brain in early and primitive primates, but has decreased greatly in size and shifted to the base of the brain in all the more highly evolved species. The cerebellum of primates, being concerned with many vital and intricate functions, has also become expanded and elaborated in its middle and lateral lobes. In the lowly treeshrews the cerebellum still lies almost entirely behind the cerebral hemispheres in a typically archaic relation, but among other primates it has come to occupy more and more a basal position underneath the occipital lobe (plate 9).

The evolutionary trend to enlarge and perfect the brain of primates has resulted in a ratio between their brain weight and body weight which is in general significantly higher than in other mammals of corresponding body size. The latter condition has to be added since it is a well-founded experience that there exists a generally valid allometric relation between body size and brain size, according to which the *relative* brain size decreases markedly with increased body size independent of other specialisations of the brain. This explains, for instance, the fact that the small, fully grown squirrel monkeys possess proportionately larger brains than do adult men, the relation between the weight of the brain and that of the entire body equalling 1 : 30 in the former, but about 1 : 51 in the latter. In a large chimpanzee of 52 kg. the corresponding ratio amounts to 1 : 119. This rule is even more clearly demonstrated by the following data appertaining to closely related animals with brains of very similar evolutionary development. The small wild cat *Felis minuta*, weighing only 1·24 kg., has a ratio of 1 : 56, in a leopard of 28 kg. the ratio has changed to 1 : 168 and in a lion of 120 kg. to 1 : 546. Since skulls are far more numerous in our museum collections than are preserved brains, we possess also many more accurate data on the volume of the brain case (commonly and simply called *capacity*) than on the actual volume of the brain. The former (in cm³) can be expressed as a percentage of the body weight (in g) with the justifiable assumptions that the specific gravity is practically one for brain and for body and that these percentages show the *comparative* relative brain sizes very reliably. If such data are plotted as curves on a scale for body weight, as in figure 52, three interesting general conclusions can be demonstrated,

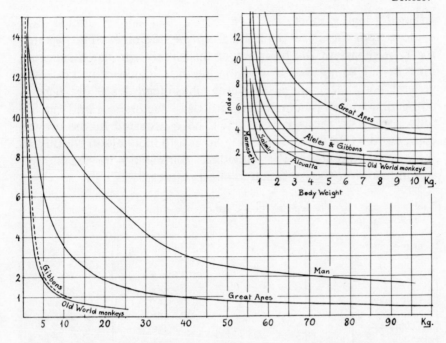

Figure 52   Curves showing the relation between body weight (in kg.) and
relative cranial capacity (capacity in cm³ in percentage of body weight in g),
constructed from data on 445 primates. Only the lower right parts of the curves
show the conditions of adults, ending with specimens of maximum body weight
(after Schultz, 1956).

namely the allometric decrease in this relative capacity with increasing body weight, the rapid initial and slower subsequent drop
in these percentages during individual growth and the various
evolutionary increases in relative brain size among the primates, as
shown by their diverging growth curves. The brain and with it the
brain part of the skull grow very rapidly during prenatal life so that
at birth the capacity has reached a much larger percentage of its
final size in adulthood than has the whole body. For instance, in
chimpanzees the cranial capacity of a newborn amounts to thirty-
four per cent of the average capacity of adults, whereas the body
weight of the former only to 3·6 per cent of the average weight of
the latter. The percentage capacity, therefore, drops in chimpanzees
from about 8·7 at birth to 0·9 at the completion of growth. By the
time the first permanent teeth erupt in simian primates (at the
approximate age of six years in man) the capacity has already

127

attained roughly ninety per cent of its final size in adults. It is evident from these few data and from the curves in figure 52 that the relative size of the brain, or of the closely correlated capacity, must be judged not only by considering the allometrically influential general body size, but also by the physiological age or stage of ontogenetic development. Only with these factors in mind can we fully appreciate the advances which have occurred in the growth curves of the relative capacities of Old World monkeys in contrast to those of marmosets, in the pongids as compared with the former, and in man over the condition in all three great apes.

Table 4   Ranges of variations and averages of the cranial capacity (in cm^3) of adult hominoids (from Schultz, 1965).

Genus	Sex	Specimens	Range of Variations	Average
Gibbon	♀	86	82–116	101
	♂	95	89–125	104
Orang-utan	♀	52	276–425	338
	♂	57	334–502	416
Chimpanzee	♀	57	282–415	350
	♂	56	292–454	381
Gorilla	♀	43	350–523	443
	♂	72	412–752	535
Man	♀	50	1129–1510	1330
	♂	92	1246–1685	1446

In its absolute size the cranial capacity varies to an extraordinary degree within each species of the hominoids. This is shown briefly by the data in table 4, which have been obtained by the writer in fully adult skulls with strictly uniform methods. Very many more capacities have been recorded by other authors, but unfortunately they are not always fully comparable, having been obtained with different techniques and in some instances the age determination seems to be open to doubt. By including all the relevant data in the literature, the ranges of individual variations would become even larger. It will be noted from table 4 that these very wide ranges of variations of the absolute capacity overlap extensively in the two sexes, and this even in orang-utans and gorillas in which the sex difference in general body size is much greater than in the other genera in the table. With such a variability of the capacity it is readily seen that the absolute size of the brain cannot be closely correlated with 'intelligence', as is often popularly assumed. These wide ranges should also serve as a warning against far-reaching speculations, based on the more or less reliably estimated cranial capacities of fossil hominids. The pronounced tendency of the brain toward individual variation is in a way also manifested by the often surprisingly marked asymmetries in the size and detailed configuration of the convolutions in the two hemispheres not only of man, but of apes, monkeys and even lemurs as well.

Chapter 10

Special Sense Organs

Eyes

IN THEIR daily activities most primates rely far more on their visual sense than on the auditory and olfactory senses, as is evident from the fact that the former has become highly perfected in various ways, while in particular the sense of smell has considerably atrophied in some prosimians and all simian primates. The predominance of seeing over hearing and smelling, which distinguishes the primates among mammals, has become possible through elaborations and refinements of the visual apparatus as well as of the relevant neural centres in the brain. With the acquisition of stereoscopic vision and the ability to distinguish colours primates can obtain far more information about their environment, near and far, than through the other special sense organs. A deaf monkey or one which has lost its power to smell might still survive fairly well, but never a blind monkey up in the trees.

In very many mammals the eyes are placed on the side of the head, so that the optic axes diverge, the visual fields overlap at most to a very limited extent, and each eye receives a different image. In primates the eyes tend to move to the front of the head and hence so much closer together that both eyes can be focused simultaneously on one and the same object, thus greatly facilitating perfect stereoscopic vision. This migration of the eyes towards the front has progressed only moderately in most prosimians, especially the long-snouted ones, but has reached its completed development in all simian primates. In tarsiers and some small platyrrhines the relatively large orbits have moved so extremely close together that the bony separation between them is at best very thin and the nasal cavity has become greatly reduced by the adjacent orbits. This evolutionary specialisation in the position of the eyes becomes recognisable only gradually in the course of prenatal life, since in early embryos the eyes appear far apart laterally and shift forward to only a slight degree during foetal development in many prosimians, but remarkably far in all other primates, including man. Binocular vision is a very essential faculty for the swift arboreal primates in judging distances before leaping and is also of the greatest help in any nearby manipulations with the prehensile and tactile fingers, typical of primates.

The ability to distinguish colours, beside light and dark, seems to be restricted among all mammals to the order of primates, as far as is known today. Colour vision, however, has reached differing degrees of perfection in various species and may be nearly lacking in some nocturnal ones, though even the lowly treeshrews enjoy a good colour sense according to recent tests. Without colour vision it would be difficult to explain the fact that the variety of colours in the skin and hair of primates far exceeds that of all other mammals. Since the gay reds, blues and yellows in the coats of many primates, often combined with contrasting colours in conspicuous designs, can hardly serve for camouflage, but certainly for mutual recognition, the monkeys themselves must be able to perceive such colours.

Many prosimians become active only at dusk or even at night and can see very well in the dark. These are chiefly the Asiatic genera *Loris*, *Nycticebus* and *Tarsius*, the African genera *Galago* and *Perodicticus* and a large proportion of the genera from Madagascar (*Microcebus*, *Lepilemur*, *Cheirogaleus*, *Avahi* and *Daubentonia*). Among simian primates there is only the one platyrrhine genus of *Aotus* which leads a nocturnal existence, possibly as a secondary adaptation. All of these primates are also quite able to see in daylight, even with considerable precision, but by the detailed construction of their retinas and other parts of the ocular apparatus they have become specially adapted to night life, that specialisation which has protected many other defenceless types of small mammals. The histological specialisations in the retinas of nocturnal primates show so many quantitative and qualitative differences among the various genera that merely the common adaptation to seeing in the dark does not necessarily imply any close interrelationships. The relevant, highly technical, microscopic findings have been excellently reviewed in a well-known book by the anatomist Clark (1959), and fully described in the large monographs of the ophthalmologists Polyak (1957) and Rohen (1962). Macroscopically the eyes of most nocturnal primates show their specialisation in exceptionally large size. This is most pronounced in *Tarsius* whose eyes form volumetrically about 4·5 per cent of the entire body and whose huge orbits are larger than its brain case (figure 2). Bushbabies, loris, aye-ayes and night-monkeys also have considerably larger eyes than diurnal primates of similar body size. On the other hand, the eyeballs of the nocturnal, but very slow pottos and *Nycticebus* show no unusual enlargement for animals of their size. In the former, particularly, the eyes are quite small, forming only 0·2 per cent of the entire body volume or weight. By having determined this relative size of the eyes in over two hundred primates the writer could demonstrate a close allometric dependency of the volume of the eyeball and of that of the orbit on general body size, according to which both become proportionately much smaller with increasing body size in all adult diurnal primates, this decrease being particularly marked among the smaller species. For instance, the same percentage relation, given above for tarsiers and pottos, amounts in macaques and gibbons of equal weight to about 0·15, but in chimpanzees and man to only about 0·03. In other words, in their relation to body size the eyes of tarsiers are 150 times larger than those of man and eyes of gibbons five times larger. This enormous decrease in the relative size of eyes (and orbits) with increasing body size is the general rule among mammals and corresponds closely to the already discussed great differences in relative brain size between related small and large animals.

The relation in size between eyeball and eyesocket differs widely among primates. In tarsiers, for example, barely half of the eye fits in the orbit, whereas in baboons and the great apes there is ample room for the eye, its muscles and a generous fatty padding within the bony chamber, which projects beyond its contents, especially on top where the heavy brow ridge provides a strong protection for the deep-sitting eyeball.

The eyelids of all primates are highly mobile and in some Old World monkeys the upper lids contrast so sharply in colour with the surrounding skin that they accentuate the eyes and in their startling motions contribute much to threatening attitudes. At the medial end of the lidcleft there persists an interesting vestige of a third lid – the so-called nictitating or semilunar membrane, which is well developed as a freely movable structure in many reptiles and birds, serving to protect and clean the exposed surface of the eyeball. Among primates the small remainder of this membrane has become entirely useless, and only in some prosimians has it been reported as being still slightly movable. In nearly all the primates investigated so far this rudiment was found to contain even a thin plate of cartilage which seems to be lacking only in orang-utans and in many, but not all, human beings.

Ears

Outer ears are a new development of mammals and have become lost again only among many of the aquatic types. In the primates the outer ears are highly variable in relative size as well as in shape, but quite uniform in the construction of their principal parts, as shown by the examples in figure 53. By measuring the height and breadth

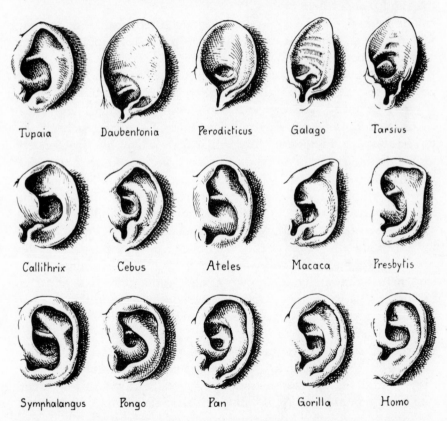

Tupaia   Daubentonia   Perodicticus   Galago   Tarsius

Callithrix   Cebus   Ateles   Macaca   Presbytis

Symphalangus   Pongo   Pan   Gorilla   Homo

Figure 53   Sketches of the outer ears of adult primates, reduced to same height and without hair. Above prosimians, middle platyrrhine and catarrhine monkeys, below hominoids.

133

of the ear and the length and total height of the head and by expressing the product of the first two measurements as a percentage of the product of the last two, one can compare the relative sizes of ears in uniformly obtained figures, which are far superior to mere words. These percentages, determined by the writer (1956) for several hundred specimens, equal only about fifteen in *Nycticebus* and *Perodicticus*, but range up to fifty in *Tarsius*, fifty-three in *Indri*, sixty-six in *Galago* and even eighty-nine in *Daubentonia*. Among all simian primates these percentages are much lower than in the great majority of prosimians, reaching at most twenty in a few monkeys and chimpanzees. Of all hominoids the latter have by far the largest ears and orang-utans the smallest of all primates with an average percentage of only three. The corresponding figure for gorilla is four and that for man five.

The central part of the outer ear, the rigid shell-like concha, serves for attachment to the head and is much less variable in size and form than the peripheral, thin and freely movable parts which flare from the rim of the concha – the scapha, helix and, whenever present, lobule. In all primates the outer ear opening is partly covered by fairly well developed hillocks – the tragus and anti-tragus – and also constantly the deep part of the concha bears on top a transverse fold – the plica principalis – which has proliferated into a sizeable flap in some prosimians, such as bushbabies and tarsiers. The large ears of the latter have several additional, but much less pronounced folds in the upper, thin part of the ear which are undoubtedly connected with the remarkable ability of these species voluntarily to fold over the free upper portion of the ear. In lower primates the edges of the unattached and movable parts of the ears are often not at all rolled in, except for a short distance above the tragus, but in many monkeys such folding over of the rim extends well across the top and in some species even to the rear. In all hominoids the edges of the ears are always strongly bent over in front and on top and very frequently also in the rear (figure 53), whereby the unattached parts become stiffened. Ears ending on top in a more or less sharp point, often called satyr-ears, are normally found only in macaques and baboons and not in any other monkeys, nor in apes. At no stage of early development are there any reliable indications of a passing recurrence of a pointed ear in man or ape. It seems very doubtful, therefore, that the so-called Darwinian tubercle, seen with differing frequency among human races, really represents an atavistic reappearance in rudimentary form of an ancient pointed ear.

The complicated shell-like shape of the outer ear is determined and maintained by the closely corresponding ear cartilage which extends to the outermost edges, except below the ear opening in the highly variable cartilage-free lobule. The latter is by no means limited to man, as is often supposed, but can be equally well developed in gorillas, chimpanzees and various Old World monkeys, whereas a true lobule has not been found in any orang-utan, gibbon, New World monkey or prosimian.

The outer ears of primates also contain small intrinsic muscles capable of modifying the shape of the upper free parts, at least in the large ears of some of the nocturnal prosimians, which can even be folded down by means of these muscles. Minute remnants of these no-longer-functioning muscles are still embedded in the ears of man, but could not be found in the extremely small ears of orang-utans. The extrinsic ear musculature is well developed in most prosimians and can also readily move the outer ears of nearly all monkeys (figure 77). Even some human beings are still capable of wiggling their ears, though merely as a parlour trick and not for discovering the direction of sound, which many mammals can do by changing the direction of their sound receptors. Tarsiers, for example, even move their right and left ears quite independently of each other in trying to locate the source of the slightest sound.

Outer ears serve primarily as funnels for catching sound waves and leading them through the auditory canal to the ear drum. This is still their main function in such large-eared nocturnal species as aye-ayes, bushbabies and tarsiers, which rely on hearing as much as on seeing. It seems more than doubtful however, that auditory acuity can generally be correlated with the wide differences in the relative size and shape of the outer ears among primates. In many mammals the thin, large ears, well supplied with blood vessels, serve as an effective temperature regulating mechanism. This is most evident in elephants, especially the large-eared African ones, and may also play a corresponding, but not very significant role in some tropical primates, such as some prosimians or chimpanzees with their exceptionally large ears. Last but not least, the ears of many monkeys can function in a mimetic capacity, since they are often strikingly abducted in threatening attitudes, whereas adducted in a relaxed mood.

The tube leading from the outer ear opening behind the tragus to the ear drum remains cartilaginous throughout life in all platyrrhines, so that in macerated adult skulls the tympanic ring and membrane are fully exposed. As had already been mentioned in chapter 2, this canal becomes extensively ossified during postnatal development in all recent catarrhines, so that in adult skulls the ear drums are hidden at the ends of bony tunnels (figure 9).

The three ear ossicles in the middle ear, which transmit vibrations of the drum to the cochlea of the inner ear, show only minor differences among primates. According to the scant information available so far, the malleus differs somewhat more than the incus and stapes, but nothing has become known about the possible functional significance of these limited variations. On the basis of the detailed morphology of these small bony levers, the treeshrews resemble certain other lower primates far more than they do any insectivores, with which they had formerly been aligned. It is of interest also that the ossicles of *Tarsius* are most similar to those of lemurs and that those of the great apes are much more like the ones of man than like those of monkeys.

The most important and complicated part of the entire apparatus of hearing, the cochlea with its spiral organ of Corti, also shows only small and as yet insufficiently studied generic differences within the order of primates. The spirals of the cochlear duct make anywhere from $1\frac{1}{2}$ to $3\frac{1}{4}$ turns, including slight intraspecific variations. Among other mammals there are some with only $\frac{1}{2}$ and others with as many as $4\frac{1}{2}$ turns, so that the range for primates is certainly not unusual. That the limits of pitch, within which sounds are audible, can differ in some primates has recently been established for marmosets, which produce and, naturally, hear high chirping sounds, imperceptible to man, except with the aid of special instruments. It is not impossible that the loud and long-sustained sounds produced by siamangs and orang-utans with the aid of distended throat pouches or by howler monkeys with their uniquely specialised laryngeal apparatus contain modulations which cannot be distinguished by the ears of other species.

Nose

While the sense of smell is of vital importance to a great many terrestrial mammals, it plays only a subordinate role in the life of arboreal primates, among which it has undergone progressive atrophy, particularly affecting the Simiae. The olfactory sense can provide information only regarding the intensity and quality of the odour of an object and, if highly developed, it can also indicate the direction of the source of an odour carried by a definite air current. Hounds, bears and some other carnivores can follow unseen prey and many ungulates can detect distant enemies by their sense of smell, but a monkey high up in the still air of a dense forest finds its mate and predators chiefly by sight, aided by hearing. Only the lower primates with comparatively little degenerated olfactory apparatus do mark their territories and can follow their kindred by the scent of special secretions. The status of female receptivity for mating can also be recognised by males of many primate species by their sense of smell, but only upon very close approach, practically to contact.

The rigid bony nasal chamber with its central sagittal partition is extended in front by an elastic cartilaginous skeleton which supports the nasal tip, largely determines its shape and prevents the nasal wings from becoming sucked in during inhalation. In prosimians these nasal cartilages are essentially two continuous tubes leading from the nasal bones and lateral edges of the bony nasal aperture to the nostrils, being separated in the rear only by the cartilaginous part of the nasal septum. In their shape these tubes vary widely, being nearly parallel in many species, such as lemurs, or sharply diverging in front, as in tarsiers, whose nostrils open sidewise. The constant distinction between 'broadnosed' platyrrhines and 'narrownosed' catarrhines actually refers to the interrelation between the nostrils, rather than to the breadth of the entire outer nose. In all the former the nostrils open toward the side and are widely separated, whereas in the latter the nostrils are directed forward and down and stand close together. The cartilaginous septum has the same narrow width in both types of noses, but the continuous cartilages, leading to the nostrils of prosimians, have become discontinuous and more or less reduced in all Simiae. A proximal part, or roof portion, is partly in platyrrhines and entirely in catarrhines separate from the distal wing portions which have thereby become freely movable (figure 8). The prenatal development of the nasal cartilages is of special interest since in the catarrhines it shows successively the stages of evolutionary specialisation in the other major groups of primates, according to the painstaking investigation of a former Chinese student of the writer, Dr. Wen (1930). Even in human embryos the nasal cartilages are still formed first as continuous paired tubes (or actually inverted gutters, since they are split below), in which there appears a lateral cleft during foetal life, exactly like the one persisting to the adult stage in platyrrhines, and only at about the time of birth is the separation between roof- and wing-cartilages completed, as a constant feature of all catarrhines of postnatal ages. It is with such apparently insignificant details that

Figure 54 Head of an adult male proboscis monkey (after a photograph of a wild animal).

138

one can demonstrate convincingly how distinguishing characters of an entire group of species appear primarily as something newly added to the processes of individual development, which initially have repeated the lasting stages of development of their less specialised relatives.

In the great majority of monkeys and apes the nose projects little beyond the general profile of the face, the nasal roots and tips are low and the wings are usually not set off from the cheeks, as they are in man, gorilla and some chimpanzees, but merge into the upper lip without any naso-labial folds. The rare exceptions to this rule of inconspicuous outer noses have appeared in three closely related Asiatic Colobinae, distinguished by the prominence of their noses in their lower, movable part. The extreme among them is represented by the proboscis monkey of Borneo. In adult males of this species the enormous, bulbous nose actually hangs over the mouth (figure 54) and, being well vascularised, can even swell slightly and become quite red during any excitement. Only the separate wing cartilages have participated in this grotesque development, while the roof- and septum-cartilages do not exceed their usual small size seen in other monkeys. No convincing reason for this unique, localised specialisation has as yet been discovered. The occasionally stated claim that it may serve as a resonating organ in the monkeys' loud calls is a most unlikely guess. It seems more reasonable to regard it as belonging in the category of decorative structures, such as the conspicuous, prominent and gaily coloured ridges on both sides of the noses of mandrills (figure 55), or the enormous cheek pads of male orang-utans (figure 12). The very fact that these puzzling and apparently useless 'decorative' specialisations become most extremely developed in mature males endorses the suspicion that they play some role in the social relations between the sexes, besides perhaps being intimidating to other animals. In female and young proboscis monkeys the noses are proportionately very much smaller and, being almost tilted upward, they resemble both sexes of the snub-nosed monkeys, *Rhinopithecus* from Southern China and the rare stump-nosed and short-tailed monkeys, *Simias* from some small islands west of Sumatra. In gorillas the lower, soft part of the nose is quite prominent, very wide and laterally well upholstered with fat (figure 13). Nevertheless the wing cartilages are much reduced and restricted to the nasal tip so that the actual wings are merely thick skin folds, unsupported by cartilage. In contrast to this

Figure 55   Head of an adult male mandrill. The entire nose is brilliant red, the lateral nasal corrugations are shiny light blue and the beard is yellowish-white.

the protruding noses of man contain extensive cartilaginous stiffening particularly in their movable wings, where the curved cartilages reach from alongside the central septum to the side of the bony aperture and closely determine the individual shape of the nose, which can influence mate selection.

The nasal cavity, halved by a central septum, contains series of scroll-like, thin, bony processes on its lateral walls and roof which form a labyrinth of so-called turbinals or conchae. Being lined throughout by mucous membrane, these processes greatly enlarge the total area of the latter and thereby also the parts which are. covered by the sensory epithelium, specialised for the reception of olfactory impulses. Mammals with a keen sense of smell generally possess an elaborate turbinal system. In elephants, for example,

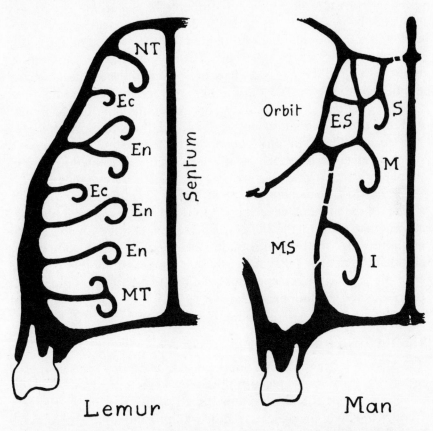

Lemur   Man

Figure 56 Diagrammatic transverse sections of the posterior part of the right nasal cavities of adult lemur and man to show the number and approximate position of the turbinals. In the lemur the large nasal chamber lies in front of the brain part of the skull and mostly even in front of the orbits, whereas in man it has shifted underneath the brain cavity and between the orbits. *NT* = nasoturbinal, *Ec* = ectoturbinals, *En* = endoturbinals, *MT* = maxilloturbinal, *S, M* and *I* = superior, median and inferior turbinals respectively, *ES* = ethmoid sinuses, *MS* = maxillary sinus.

there are numerous primary and accessory, extensively convoluted turbinals which by far surpass the corresponding conditions in even the most primitive primates. Among the latter the aye-aye has five endo- and at least three ecto-turbinals, as well as a comparatively large olfactory bulb, indicating that this nocturnal animal still relies somewhat more on its sense of smell than do the other recent prosimians. In contrast to the turbinal system of treeshrews and lemurs with their proportionately large nasal cavities (figure 56), that of *Tarsius* and those of all simian primates have become extensively reduced. In tarsiers this is undoubtedly due to the extreme diminution of the nasal chamber through the unique enlargement of the orbits, but all monkeys, apes and man have shared a clear trend to reduce the number of conchae and to withdraw the olfactory epithelium to the remotest uppermost region. In man and the great apes, for example, only an inferior (= maxilloturbinal) and a middle and superior (= endoturbinals I and II) conchae are normally present in adults, with only occasionally one more, or even one less. The so-called nasoturbinal is still fairly well represented in various monkeys and in gibbons, but has shrunk to a tiny remnant (= *agger nasi*) in man and the great apes. Through several careful studies of the prenatal development of the turbinal system it has been shown that early in the life of man there are still as many as five ethmoturbinals laid down, besides the maxillo- and nasoturbinals, and that the marked postnatal reduction represents another beautiful example of the direct and close connection between ontogenetic processes and evolutionary changes.

Tactile Structures

The skin of primates is well supplied with sensory nerves as special receptors capable of feeling pressure, heat and pain, but this sensitivity is not equally developed in all bodily regions. At the very front of the body the nostrils of nearly all prosimians are surrounded by hairless, glandular and hence moist skin, which is highly sensitised and thus acting as a tactile organ. This specialised skin, found also in a great variety of other mammals, develops from the early embryonic median nasal process of the head and reaches in the Tupaiiformes and Lemuriformes all the way to the mouth (figures 3 and 4). In some of the Lorisiformes this so-called rhinarium has begun to become limited to the nasal region, since the embryonic, lateral, maxillary processes meet one another superficially in the labial region, producing uninterrupted hairy skin on the upper lip. In the tarsiers and all simian primates, the rhinarium has disappeared entirely through the complete fusion of the maxillary processes in this region, involving also the muscularisation of the upper lip across the midline, which in turn has improved its movability and thereby facial expression.

Many mammals possess on their forearms and, particularly, their faces specialised tactile hairs, or vibrissae, implanted by large and well innervated roots in cutaneous hillocks, which transmit impulses from contact with their environment or even from the air repulsed by the mere approach to a firm surface. These long and stiff vibrissae serve on the brows for the protection of the eyes as a warning against nearby obstructions, and the carpal vibrissae, which reach as long levers towards the palms, register the approach of a landing place after a leap and release the grasping reflex in the fingers within a fraction of a second. Treeshrews and lemurs (figures 3 and 4) are still supplied with practically all these typical tactile hairs on the brows, cheeks, lips, chin and distal parts of the forearms, but in the lorisiform prosimians they have become considerably reduced, indeed, usually even more so than in the tarsiers. Among the simian primates the tactile hairs have undergone further reduction, being still fairly well developed on the brows and forearms in marmosets and on the brows alone in some other platyrrhines, whereas in all Old World monkeys they are at best very inconspicuous and without functional significance and in all hominoids they have disappeared entirely in adults. It is interesting to note that the phylogenetic degeneration of these tactile organs, which is most advanced in catarrhines, has not eliminated them completely in ontogeny. For instance, in all foetuses of guereza monkeys, studied by the author, there are still from two to four long vibrissae implanted in a hillock on the ulnar side of the forearm close to the wrist and pointing toward the palm, but no traces of them could be found in juvenile or adult guerezas. Even in a number of early human foetuses the carpal hillocks have been found in their typical place by the Swedish embryologist Broman as well as by the writer, but no vibrissae have ever been seen to persist to later stages of human development.

Much more delicate, varied and informative tactile sensations than those perceived through vibrissae are obtained by primates by

143

means of the touch-pads on their palms and soles and the cutaneous ridges which develop on and between these pads. These specialised parts of the skin are densely supplied with sensory corpuscles and form an admirable mechanism for tactile discrimination which has become best perfected in the higher species as a vital addition to the usefulness of their hands. It has already been mentioned that comparable tactile specialisations of the skin exist also in the nasal region of prosimians and on the ventral side of the prehensile tails of some platyrrhines. Simple, but undoubtedly tactile cutaneous tubercles and ridges, corresponding to the much more numerous and complicated dermatoglyphics on the volar sides of the hands and feet, have developed also on the rhinarium of bushbabies, on the ventral side of the tail of one species of tarsier and on the dorsal sides of the hairless middle and terminal phalanges of the fingers of gorillas and chimpanzees. More about the dermatoglyphics of primates has been discussed in the chapter on the specialisations of the skin, since these structures serve not only in connection with the sense of touch, but have also an important function as friction skin.

Chapter 11

Growth and Development

INFORMATION regarding the profound changes with age in the size and differentiation of the body is of basic importance for the understanding of the appearance and nature of all specialisations. Age changes really occur throughout life, but they are most dramatic by far during the initial intrauterine period, to proceed with generally decreasing tempo until the comparatively stable conditions of adulthood have been attained. Though this book is concerned chiefly with the life of adult primates, it must be emphasised that the distinguishing characters of adults are merely the temporary results of an intricate preceding chain of inherited processes of individual development. All evolutionary innovations acquired by adults are primarily due to some alteration in at least one of the complicated details of growth and development. Specialisations, therefore, do not become fully marked as a rule until rather late in development, just as they were not yet present in the earlier evolutionary stages of the species. This intriguingly close relation between individual and evolutionary history had led to the formulation of the old and rather crude biogenetic law, later modified into the widely accepted recapitulation theory, which bluntly claims that ontogeny (= individual development) repeats phylogeny (= evolutionary development) in a rapid and more or less abbreviated manner. It seems preferable and in much closer agreement with the more recent accumulation of knowledge to see this fundamental problem in a different light by stating simply that phylogenetic changes are directly connected with modified ontogeny. It is from this point of view that the examples of age changes in primates will here be presented. For instance, at early stages of prenatal life man and monkey are still strikingly alike, as shown by plate 8. Both these primates possess proportionately huge heads, stout trunks, very short limbs with broad hands and feet, small ears and as yet hardly any eyelids. Naturally, nobody could assert that these are typical features of their common ancestors in adult life, but it can be assumed with the greatest confidence that the embryos of their common progenitors had also looked about like this and that the final specific characters diverged more and more under the progressing influence of differences in their modified ontogenetic programmes.

It is customary to distinguish between mostly quantitative changes, due to growth, and chiefly qualitative ones, as representing development, but these terms cannot be used consistently in their literal sense. Thus, 'growth' includes not only increases in absolute size, but also any partial decreases with advancing age and is concerned particularly with the significant changes in relative size, due to differential rates of growth in the various parts of the body. Similarly, 'development' does not imply only increased differentiation, but refers also to the gradual degeneration and loss of such characters as are normally limited to transitory stages of ontogeny.

The duration and intensity of growth in general differ widely among primates between such extremes as *Microcebus*, which becomes fully grown in less than a year with a body weight of only about 60 grammes and male gorillas which can reach a weight of even 200 kg. in about eleven years. The tempo of the increase in body weight diminishes rapidly with advancing age. In man, for example, the microscopic fertilised ovum, weighing about 0·005 milligrammes, grows during the eight weeks of embryonic life to roughly 1·1 grammes, or per week 27,500 times its initial amount. During the following foetal period man attains his normal birth weight of about 3,200 grammes at the approximate average weekly rate of ninety times the beginning weight and during postnatal growth this rate per week dwindles to a mere 0·02 times the weight at birth. Comparable data for nonhuman primates are unfortunately still very scarce, but they certainly agree with the rapid decrease in the speed of growth which can be amply demonstrated for man, and leave no doubt that the rate of growth per unit of time differs not only in the various periods of life, but also according to species and even according to sex in some species.

The duration of the different periods of growth shows among primates a general trend of prolongation with evolutionary progress. Prenatal growth lasts six weeks in treeshrews, nine weeks in *Microcebus*, ten weeks in *Cheirogaleus*, about eighteen weeks in lemurs and up to twenty weeks in bushbabies, or even in prosimians generally much more than in insectivores and rodents of comparable sizes. The gestation period of marmosets as well as of the other platyrrhines seems to fluctuate but little around twenty to twenty-five weeks according to the still very modest number of reliable observations gained so far. Macaques and baboons are born at the accurately known average ages of twenty-four and twenty-seven weeks respectively. In gibbons the corresponding age has advanced to about thirty weeks and in the great apes to even thirty-four weeks on an average of many records for chimpanzees and to possibly as much as thirty-eight weeks in orang-utans as well as gorillas according to the few relevant reports published so far. The latter figure equals the average 'conception age' of human newborns. In spite of the fact that gestation has been most prolonged in the largest recent primates, there is no close or consistent correlation between body size and duration of prenatal growth, as has been claimed, since gibbons, for instance, are much smaller than baboons yet have a longer pregnancy, and spider monkeys and marmosets differ widely in size, but hardly at all in their gestation periods. As will be shown later, the state of development or maturation at birth does not automatically increase with the duration of prenatal life, but seems to be correlated with the parental care and protection the offspring can expect in the different types of primates. The termination of intrauterine growth must necessarily also be in close harmony with the relation in size between the full-term foetus and the maternal birth canal, as will be discussed below.

The evolutionary trend among primates to prolong their main periods of life is most evident postnatally, as shown by the accompanying diagram (figure 57) which must suffice here in place of elaborate tables of detailed data, which could be collected from a widely scattered literature. The period of infantile life, which terminates with the appearance of the first permanent teeth according to its customary anatomical definition, lasts in prosimians for only a fraction of its duration in monkeys and in the latter never as long as in the highest primates, among which it has become prolonged least in gibbons and by far the most in man. The second postnatal period, the juvenile one, extends from the age of eruption of the first to that of the last permanent teeth, when adulthood is attained according to the desirable aim, clearly to define comparable stages of development. This is the period during which the delicate milk teeth become replaced and added to by larger ones, while general growth proceeds at a slower rate than before until the growth curve changes its rise to a practically level direction. The duration of juvenile growth amounts to only a few months in treeshrews and up to at most eighteen months in other prosimians. It has increased to somewhere between two and three years in monkeys and five years in gibbons. In all three great apes this period lasts on an average eight years and in at least some human races even fourteen years. In macaques the dentition is completed and noteworthy growth has practically ceased at the approximate age of seven years, in the great apes at eleven years and in man at twenty years. This refers to males, since in a good many species postnatal growth as well as development can proceed appreciably more rapidly in females, as is well known for man.

This clear and partially very marked trend for the prolongation of youth and immaturity has profoundly influenced the behaviour of primates in various respects. The prolonged close dependency of the young provides correspondingly increased possibilities for observing and copying the actions of adults, i.e., for learned rather than instinctive behaviour. On the other hand, the more and more retarded maturity postpones the addition of new generations, which is so vital for the maintenance of species limited mostly to single offspring at a birth.

During the final and longest main period of life, rather crudely held together under the term adulthood, bodily changes continue at such slow rates generally that they are not readily recognised. Especially in the lower primates there still can occur considerable localised growth in the length of the limbs, the girth of the chest, etc. in connection with the as yet incomplete skeletal development at the age of finished dental eruption. Toward the close of the adult period the body again undergoes more pronounced age changes, culminating in senile degeneration. In this period of decline, rare and brief in wild primates, some tissues tend to decrease their elasticity, bones atrophy and fracture easily, the misnamed permanent teeth, extensively worn, break down and become lost altogether, while the animal loses in weight.

Figure 57   Diagrammatic representation of the approximate durations of the main periods of life in some primates (after Schultz, 1956, revised according to the latest information). Infantile period to appearance of first permanent teeth, juvenile period to completion of permanent dentition, adult period to end of life-span in a majority of healthy individuals.

17 Old male eastern gorilla leading five others in forest.

18 Cotton-headed tamarin.

19 White-fronted marmosets grooming.

20 Young male Bonobo (*Pan paniscus*).

21 Infant gibbon of three weeks clinging to mother (*Hylobates lar*).

22 A family of gelada baboons (*Theropithecus*)
23 Marmosets (*Leontocebus rosalia*) sitting upright while feeding with the aid of their hands.

24 Rolloway monkeys (*Cercopithecus diana*) with white capes and beards.
25 Adult potto (*Perodicticus*).

26 Angwantibo (*Arctocebus*) with infant.
27 Dwarf galago (*Galago deomidovii*) threatening.

28 Slender loris (*Loris tardigradus*) with infant
29 A family of night monkeys (*Aotus*)

30 Adult Mindanao tarsier standing upright.

31 Adult male Goeldi's tamarin (*Callimico*).

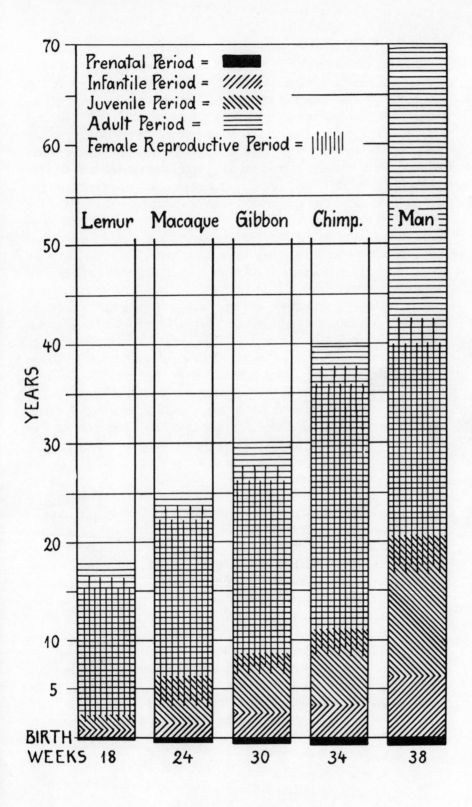

That the total duration of all these periods – the average life span – has also increased intensively with progressing primate evolution is clearly evident from the examples in figure 57. The longevity of primates in their natural environment is unfortunately unknown and can only be estimated on the basis of records of specimens which have lived to their death in the protection of captivity. The latter has become quite efficient in modern zoos and laboratories with careful medical supervision, balanced diet and absence of major excitements and overexertion. For this reason many references can be found in the newer literature on apes and even some monkeys which have lived in captivity to well over thirty years, but such instances are exceptional. Nevertheless they have been taken into consideration in estimating the average life spans in our diagram, which may represent optimal, rather than 'natural' values. It is of interest to note that the total duration of life equals roughly $3\frac{1}{2}$ times as much as the combined infantile and juvenile periods in at least all four of the different catarrhines in figure 57. The average human life span of seventy years is certainly unique among primates and has been reached only in very recent times, supposedly with and due to modern civilisation. Since this maximum of absolute duration has the same relation to the length of the preadult period as in many nonhuman primates, one wonders whether the latter period, too, may have increased so much only in recent man, especially in view of the fact that its present duration of twenty years also represents an outstanding extreme.

Of greatest importance during individual development is the onset of sexual maturity and the duration of the period of fertility with the accompanying endocrine changes which influence growth, behaviour and the formation of secondary sex characters. Even though these topics will also be referred to in later chapters, they are briefly discussed here in connection with the diagram, showing the duration of the main life periods. Female treeshrews become sexually mature in the middle of the fourth month and can produce a new generation before they are half a year old. Among the more advanced and larger prosimians the age of beginning sexual maturity in females varies widely and in some forms may not occur before the end of the second year, but at any rate much sooner than in the simian primates. The ages of reproduction in the latter have become best known in macaques, which are kept in large colonies for the study of their sexual physiology. In these and some other catarrhines the female sex cycles are readily recognised by periodic external swellings which are correlated with ovulation and menstruation. These cycles appear some time before an actual impregnation can take place, indicating that menarche is followed by a period of adolescent sterility in monkeys, as it does in man, and only then begins the period of female fertility, which in most species starts well before the close of the juvenile period. The absolute ages, at which first impregnations can take place, vary considerably, but follow closely the increases in the ages of maturation in general, as found among the primates. This implies that even the shortest possible interval between succeeding generations has become greatly lengthened with evolutionary progress and that species with advanced ages of beginning fertility will have fewer generations and generally less offspring during a given span of time than species with earlier ages of reproduction. Furthermore the speed of population replacement and of potential evolutionary change is very much slower in the former than in the latter species. As indicated in the diagram of figure 57, the length of the entire reproductive period of females has not increased as much as has the life span in all the types shown, since it is comparatively short in man, in whom the menopause occurs long before the end of the average longevity. Though it is not yet an established fact, it does seem highly probable that healthy female monkeys and apes may be able to conceive as long as they live and have no final period of sterility, as in human females. At any rate there is a growing number of cases of captive female monkeys and apes which become pregnant at ages close to the maximum longevity records for their species. In all these cases, therefore, fertility lasted about as long in females as in males, instead of ceasing much sooner, as is normal in the human species.

During the entire period of fertility the primate mother has to support repeatedly the parasitic growth of her unborn child. The final weight of the latter (without the additional weight of the afterbirth, the foetal membranes, cord and fluid) can reach at birth up to much more than ten per cent of the non-pregnant maternal weight according to the data in table 5 and if it is taken into consideration that species, such as those of *Tupaia*, *Microcebus* and *Callithrix*, regularly have multiple births, while the figures in the table refer only to single ones. The smallest of all these relative birth weights are those of the great apes. It is very puzzling that the largest of the nonhuman primates should produce such small newborns. For

Table 5 Average percentage relations between the weights of newborns and of adult not gravid females of corresponding species according to records of the author and additional data in recent literature.

Species	Percentage	Species	Percentage
Tupaia glis	7·0	Papio hamadryas	7·1
Microcebus murinus	6·0	Theropithecus gelada	10·0
Lemur fulvus	4·6	Presbytis cristatus	7·0
Galago senegalensis	5·0	Presbytis entellus	6·7
		Nasalis larvatus	4·6
Callithrix jacchus	12·1		
Saimiri sciureus	14·7	Hylobates lar	7·5
Cebus capucinus	8·5	Siamang	6·0
Alouatta palliata	8·0	Orang-utan	4·0
Ateles geoffroyi	7·0	Chimpanzee	4·0
		Gorilla	2·6
Macaca irus	10·0	Man (white)	5·5
Macaca mulatta	6·7		

instance, the viable gorilla, born in the zoo of Columbus, Ohio, weighed only 1·47 kg. and the first gorilla, born in the zoo of Bâle to a young mother of 70 kg., weighed 1·82 kg. A fine chimpanzee, born in the writer's former colony of these apes, weighed 1·41 kg. and the many others born elsewhere have similarly low weights. The recorded birth weights of four orang-utans vary between 1·4 and 1·6 kg., all having been perfectly normal and viable. The average weight of the human newborn is usually given as at least 3·2 kg., or far more than the corresponding figures for any other primate. This discrepancy is partly due to the fact that most human babies are born well padded with a remarkable amount of subcutaneous fat, whereas monkeys and apes have very little, so that they look decidedly 'skinny' and horribly wrinkled (figure 58). Delivery has to take place at a time before the growing foetus has become too large to pass the maternal birth canal. This vital relation depends chiefly on the size of the head, which is comparatively large in all newborn primates and approaches the diameters of the pelvic passage so closely in most species that the actual birth is often a very prolonged and difficult affair and can even end fatally. From the Röntgen picture of a macaque shortly before delivery (plate 10) it is readily seen that there is no room to spare during the passage of the foetal head through the pelvic ring. Fortunately the bones of the former can be slightly telescoped at their open sutures without permanent harm and the width between the hip bones can be appreciably

152

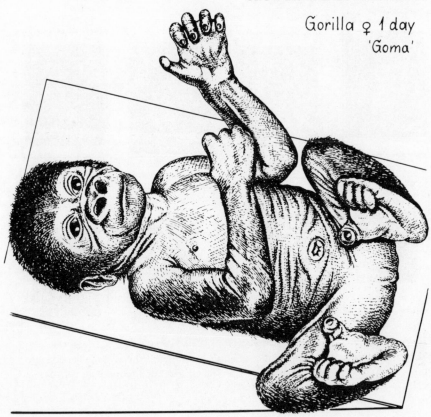

Gorilla ♀ 1 day
'Goma'

Figure 58   Captive-born female western gorilla, one day old (after photograph by E. M. Lang, 1959, in Documenta Geigy, Bull. Nr. 1, Basel).

increased through stretching of the tissues in the sacro-iliac and symphyseal joints. Monkey babies are normally born head first, so that there is no danger of the umbilical cord becoming clamped off between foetus and pelvis before breathing can begin. The head is turned with its largest diameter passing between the maternal promontory and pubic symphysis, which in nonhuman primates is always the longer of the two diameters of the birth canal. In man alone has the sacrum become pushed down to a position opposite the pubic symphysis along its entire extent and the sagittal diameter of the pelvic inlet is usually slightly smaller than the transverse one, so that the foetal head is often turned sideways for a less difficult passage. These conditions are also indicated in the diagram of

153

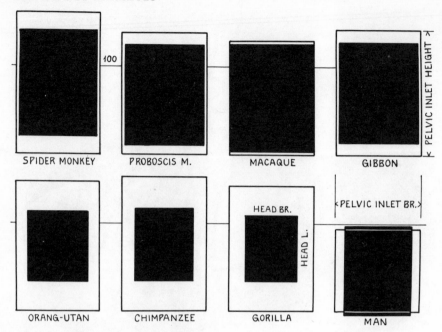

Figure 59   Diagrammatic representation of the relation in size between the average diameters of the pelvic inlet of adult females and the average head length and breadth of newborns of the same species, all diagrams reduced to the same pelvic inlet breadth (after Schultz, 1960).

figure 59, from which it is at once apparent that the three great apes are very exceptional as far as the relation in size between their birth canals and newborns is concerned. In confirmation of this finding it is commonly observed that in these apes the actual births occur speedily and rarely with noteworthy difficulty. This forms a considerable contrast to the average conditions in man and in most of the monkeys whose births processes have been witnessed. With such exceptionally and 'unnecessarily' wide birth canals the great apes could theoretically afford to continue their already prolonged intrauterine development until the foetuses have attained a relative size and a state of maturation comparable to that prevailing among monkeys. Actually and regardless of their conditions for easy delivery the apes are born as helpless and immature as is the exceptionally large human newborn.

The size of the full-term foetus is undoubtedly the most decisive factor determining the end of prenatal life. Of further great significance is the stage of maturation reached during intrauterine development, which in turn determines the degree of helplessness of the infant. Among the manifold processes of general development the state of skeletal ossification can be accurately assessed by x-ray photography and forms one of the best means for measuring the progress toward maturity. The drawing of a stained and cleared monkey foetus in plate 11 exemplifies an early stage of skeletal development in which many peripheral parts and the ends of long bones are still entirely cartilaginous (showing the dark stain in the figure) and only in the central parts of most skeletal elements has the primary cartilage been replaced by bone. Most of the many separate parts of the skull are still paper-thin and those forming the cranial vault are as yet a mere shell, incompletely covering the rapidly growing brain. These vault bones develop directly in membrane from centres of ossification from which bone radiates in all directions, leaving the remotest corners unossified to the last as the so-called fontanelles, clearly shown in the drawing. At birth these fontanelles have nearly or entirely disappeared in most monkeys and have usually become quite small in apes, but in man they are still good-sized and the great fontanelle at the vertex remains open for a considerable time after birth. The long bones of the limbs of new-born monkeys have already developed centres of ossification in most of their cartilaginous ends – the so-called epiphyses – and in at least a majority of their carpal and tarsal elements as well as in the epiphyses of some of their digits. As is shown by the examples in figure 60, there are not nearly as many centres of ossification in the newborns of the great apes as in those of most Old World monkeys and the average human newborn has advanced even slightly less in skeletal maturation. Individually some viable full-term foetuses of great apes show no more progress of ossification than do occasional human babies at birth. The precise stage of ossification of a typical newborn macaque is not reached by apes and man until the second half of their infantile life or even later. The progress in the development of the dentition, the hair and other morphological features, described later, is also far more similar at birth in man and the great apes than in the latter and monkeys. The same can be said in regard to the physiological and behavioural development at and after birth, which has been studied chiefly in recent years in different representatives of catarrhine primates. For instance, while bushbabies actually begin to walk on all fours within a few hours after they have

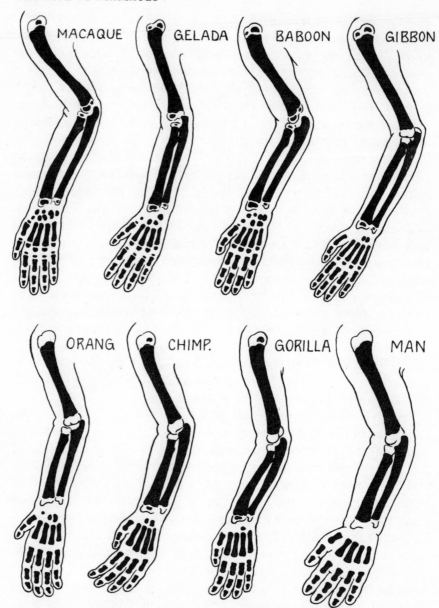

Figure 60   Tracings of x-ray photographs of the upper extremities of newborn catarrhine primates, reduced to approximately equal total length.

been born, some guenons after only two days, and macaques usually toward the end of the first week, chimpanzees require twenty weeks for this stage of their postural development, the few gorillas on record slightly more, and in the slow postnatal maturation of man the ability to 'walk' in a fashion is not reached until about forty weeks. All such facts flatly contradict the frequently heard vague claim that man is unique in his being born utterly helpless in such a very immature state as is very exceptional among primates. Such prosimians as *Tupaia*, *Microcebus* and *Cheirogaleus* are born blind, practically naked and quite unable to hang on to anything, so that the mothers have to carry them in their teeth.

In spite of man's extraordinary size at birth it is soon overtaken by that of the great apes, owing to marked differences in the rates of postnatal growth. In body weight male gorillas begin to surpass the average weight curve for man in the third year and even the much smaller chimpanzee reaches the human growth curve with four years and remains above it throughout its juvenile period, but after the age of about eleven years the weight of man continues to rise for a long time, while that of the chimpanzee has ceased to rise to a significant extent. Owing to the wide interspecific differences in the growth rates of the various parts of the body the typical body

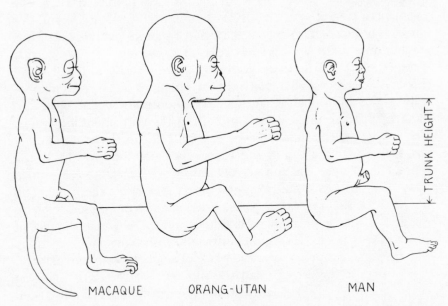

MACAQUE   ORANG-UTAN   MAN

TRUNK HEIGHT

Figure 61   The body proportions of a newborn macaque, orang-utan and man, reduced to same anterior trunk length (after Schultz, 1926).

proportions of adults develop only gradually and are still far more alike in early than in late stages of growth. This important fact, to which attention has already been called with plate 8, is here emphasised again with the few examples in figure 61, redrawn from one of the writer's earliest studies of these conditions. It is readily seen that at birth the monkey, man and ape have still much more similar proportions than they finally acquire when adult. In all three examples the head, particularly its brain part, is relatively large, the trunk is not yet nearly as slender in the monkey nor as plump in the ape as later on, and the proportionate lengths of the limbs are not yet so far apart as when their growth has been completed. It is seen specifically that in relation to the length of the trunk man is not distinguished among newborns by the largest diameters of his head nor by the great length of his lower limbs. An endless variety of other proportions, which become very different in adult primates, can be explained as primarily due to differing rates of localised growth. For instance, in early human foetuses the middle toe is still the longest and the first toe very short, just as in nearly all other catarrhines, but subsequently the former toe increases less and the latter more to produce the changed digital relations characterising the adult human foot (plate 6). As will be discussed later, the various tailless primates all possess an outer tail early in life, but this appendage grows so slowly that it soon becomes overgrown by the surrounding parts of the pelvic region. During the ontogenetic changes of the skull there gradually appear many detailed topographic differences, characterising the adults of the various taxonomic groups of primates. Most of these differences can be traced back to differing intensities of growth in the corresponding cranial elements. To mention merely one example, illustrated by the skull of the foetus in plate 11, of the four bones surrounding the lateral fontanelle, the parietal and narrow alisphenoid usually meet in man soon after birth and thus prevent a junction between the frontal and temporal bones. The latter two, however, grow more intensively in the great majority of monkeys and apes and thereby gain the race of all four elements in closing the fontanelle with bone and in forming a later fronto-temporal suture in place of a spheno-parietal one.

The *ages* of appearance and the numbers of certain ossification centres during early skeletal maturation have already been discussed with reference to the marked retardation of these developmental processes in apes and man in contrast to monkeys. The general *sequence* of appearance of such centres seems to be much more similar than are the ages of appearance in all the types studied so far. Even more constant is the detailed order of the later fusion of the epiphyses with the shafts of the long bones toward the end of juvenile or early in adult life, when noteworthy growth in the length of these bones has ceased. According to all the detailed data for this process, this order of bony fusion between the shafts and ends of the limb bones begins at the elbow and proceeds to the hip, the ankle and the knee and ends at the wrist and shoulder in lemurs, marmosets, spider monkeys, various Old World monkeys, all man-like apes and man with only very few and quite minor exceptions. The period of this tenaciously maintained programme of epiphyseal union has changed, however, in its relation to the final stages of dental development.

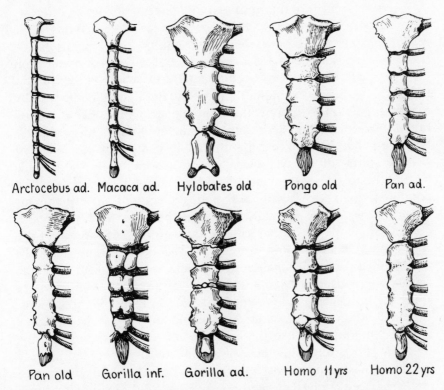

Figure 62   The sternum of some primates, reduced to same length. Note different numbers of ribs reaching sternum and lack of fusion between intercostal segments in the adult prosimian and monkey in contrast to gradual fusion with advancing age in apes and man.

Among all the lower primates merely the first few epiphyses have become fused at the age of completion of the permanent dentition, when the great majority of the epiphyses are still separate. In orang-utans and even more so in chimpanzees and gorillas the period of epiphyseal closure has shifted so that less of this skeletal development remains unfinished by the time the last teeth have erupted. In some human races only the epiphyses at the wrist and shoulder are still separate when the dentition has been completed and in Europeans the extreme shift has been reached with the simultaneous ending of both these periods of skeletal and dental development. This is merely one of many similar stories showing that there can occur both accelerations and retardations in the sequence of age changes without necessarily affecting the final appearance of the fully grown body.

The development of the sternum of primates provides an interesting example of the direct connection between altered ontogenetic processes and phylogenetic changes in adults. In all primates the bony sternum develops from a single or double series of ossification centres, appearing in the spaces between the insertions of the ribs. As indicated by the few examples in figure 62, these serial bones usually remain separate throughout life in prosimians and monkeys, whose breast bones are somewhat flexible and are joined by anywhere between seven and twelve pairs of ribs. In contrast to the long and slender sterna of all the lower species, those of the hominoids become fused with advancing age at least in the part known as the corpus sterni, which finally turns into a single solid bone. This fusion of sternebrae normally begins at the lower end, progresses upward and ends not infrequently even with the coalescence of the corpus and the uppermost segment (= manubrium) in old age. In the Hylobatidae and Pongidae the process of solidification of the corpus sterni usually begins with and continues through most of adult life, to become fully completed only in really old specimens. In man, however, the same ontogenetic innovation has become accelerated and shifted to the later part of juvenile life. Thus, the sternum of a barely adult man of around twenty years has already become almost indistinguishable from that of a very old chimpanzee (see lower left and right pictures in figure 62). Incidentally, the breast bones of the hominoids are also so much broader than those of the lower primates that the former group had sometimes been referred to as 'Latisternalia'. This very striking distinction is partly due to the fact that in hominoids there are often paired ossification centres in each sternebra and usually only seven or even six pairs of ribs reach the sternum, fewer than seven being particularly common in orang-utans and siamangs, which have the shortest and widest sterna. The latter and their close relations, the gibbons, are further specialised by having their manubrium sterni commonly extend over the first two intercostal spaces through an accelerated fusion of the corresponding ossification centres.

How helpful ontogenetic findings can be in evaluating specialisations of adults is evident from the developmental conditions of the wrist bone. The latter contain a separate central element – the os centrale – in a great many different vertebrates, including most prosimians and all simian primates, except adults of chimpanzees, gorillas and man. In the latter, as in the former, however, a centrale develops early in prenatal life, only to fuse with the adjoining navicular element of the carpus in the three exceptional higher primates. This disappearance of a separate centrale through fusion has become most accelerated in man, in whom it occurs normally during the third month of intrauterine development and only very rarely remains independent. In chimpanzees and gorillas an identical fusion takes place much later, varying individually between the last part of foetal life and occasionally even as late as the first part of juvenile life. In orang-utans, siamangs and gibbons, finally, a corresponding fusion has been found repeatedly, but only in very old specimens. An os centrale, therefore, is not simply a qualitative character, differentiating primates by its presence or absence, but primarily it is the result of a modification of an ontogenetic process in regard to the relative age at which it loses its separate existence. Neither the strength nor the flexibility of the carpal skeleton is significantly changed by the fusion of the small centrale with the large navicular bone and hence it can have no selective value. Yet an identical trend for this particular fusion has appeared in all hominoids in very different degrees as one of many manifestations of their common origin and subsequent divergence.

Of all changes of the skeleton with age those of the skull have been most intensively studied and contain many conditions of significant help for the proper evaluation of the manifold specialisations among the skulls of adult primates. This close relation between cranial ontogeny and phylogeny is best illustrated by the following few examples. As in many other mammals the two halves of the lower jaw do not fuse in front in nearly all prosimians. In the simian primates, however, the soft tissue, which initially holds the mandibular sides together at their symphysis just as in prosimians throughout life, becomes replaced by bone during the infantile period and thereafter the entire lower jaw acts as one solid lever in consequence of this slight addition to cranial development, acquired without exception by all simians. In a similar way and as has already been mentioned, an osseous auditory canal, lacking in platyrrhines, appears and grows after birth in all catarrhines as an ontogenetic innovation, producing a phylogenetic specialisation in adults.

The cranial sutures generally become obliterated sooner or later during adulthood and some even much earlier. It is at these sutures that much of the peripheral growth of bone takes place and with sutural closure there is practically no more increase in the cranial diameters perpendicular to a particular suture. This is very evident from the changes in size and shape of human and ape skulls with abnormally early or late obliteration of one or another suture, especially of the cranial vault. The relative age and the sequence of the closure of the many sutures can differ markedly according to the species or genus, but is fairly regular intraspecifically. As a rule most of the sutures disappear very soon after the attainment of adulthood in the great apes and some other catarrhines in striking contrast to these conditions in man, whose sutures normally remain open much longer. This comparative retardation in man is hardly a special adaptation for the exceptional growth of his brain, as it has often been interpreted, inasmuch as the sutures of most of the New World monkeys remain open until an even older relative age than prevails in recent man.

For a long time it had been claimed that man differed radically from all other primates in lacking intermaxillary bones (premaxillae), also called the os incisivum since in it develop the upper incisor teeth. After the 'discovery' of the human intermaxillary, for which the famous German poet-scientist Goethe is commonly but incorrectly given credit, it was realised that actually man is distinguished only by the absence of sutures between the intermaxillary and maxillary bones of the anterior surface on the facial part of the skull. The borders between these bones, which later form sutures in all nonhuman primates, disappear early in the development of man by becoming covered by thin outgrowths from the maxillaries. This very localised deviation from the usual order of sutural changes is no more remarkable than many other minor mutations, limited to one or another species. For instance, the suture between the two nasal bones closes before birth in macaques and chimpanzees, but usually remains open into adult life in most other primates; indeed in man it is rarely obliterated even in senile individuals. The particular trend toward an early disappearance of the facial intermaxillary sutures has affected not only man, but also the pongids inasmuch as these sutures begin to close in nearly all infantile apes and have usually become entirely obliterated at much earlier relative ages than in monkeys. Such detailed ontogenetic findings, trivial as they may appear to the layman, are as helpful for the student of primate evolution as are morphological similarities which are found in adults.

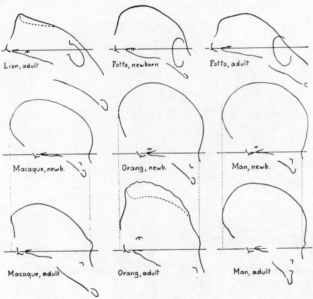

Figure 63 Midsagittal sections of the skulls of some newborn and adult primates and of an adult lion, oriented according to the axis of the skull-base (*nasion – basion*) and reduced to same length of brain-case.

In all mammals, except the suborder of simian primates and tarsiers, the occipital condyles of the cranio-vertebral joint lie at the very end of the skull so that the entire weight of the head must be carried in front of the spinal column. This extreme position of the condyles at the rear of the skull is found at all stages of development in rodents, ungulates, carnivores, etc. as well as in prosimians (figure 63). In all monkeys, apes and man, on the other hand, the occipital joint has shifted far forward to a position underneath the skull, where it remains through foetal development. During infantile and early juvenile life the skull grows more intensively in its pre-condylar than in its postcondylar parts in all monkeys and apes, so that their condyles migrate to the rear in varying degrees before they have become adult, as is shown by the examples in the figures 39, 40, 64 and 65. In man the nearly central position of the cranio-vertebral joint, typical of all simian foetuses, remains practically unchanged throughout growth and hence the human head stays balanced on top of the spine with its centre of gravity only a little in front of its support at the occipital condyles. In contrast to this

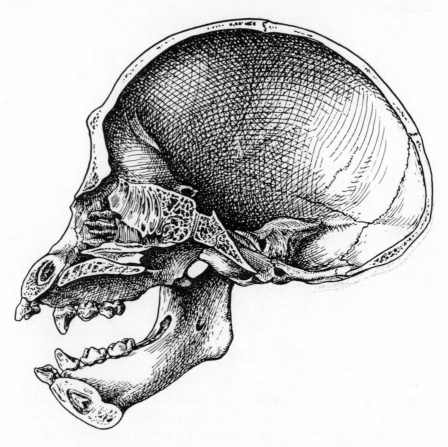

Figure 64   Midsagittal section of the skull of an infantile chimpanzee.

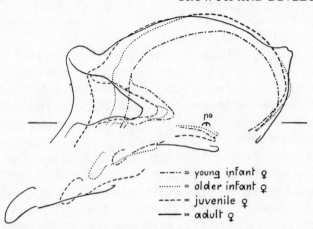

----·--- = young infant ♀
··········· = older infant ♀
---- = juvenile ♀
—— = adult ♀

Figure 65   Superimposed midsagittal sections and sagittal sections through middle of orbit of four skulls of gorillas of different ages, reduced the same amount and oriented in ear eye horizon with coinciding ear opening (= *po*). The approximate centres of the occipital condyles fall along one perpendicular line (after Schultz, 1956).

human peculiarity, which is primarily an ontogenetic one, we find all degrees of condylar migration among the other Simiae, least pronounced in squirrel monkeys and *Pan paniscus* and most in such species as howler monkeys, baboons, orang-utans and gorillas. The outstanding human specialisation in the position of the adult head in relation to the spinal column is commonly regarded as a necessary adaptation to erect posture and is also often quoted as the main support for the theory of man's unique retention of foetal conditions, which will be discussed later.

As is shown by the figures of sagittally sectioned skulls, the face part is proportionately much smaller when compared with the brain part, in early than in later stages of growth and does not yet project nearly as far forward in the young as in the adult. At the same time the orbits move during growth from their initial position entirely underneath the brain cavity to one largely in front of it, leaving in adults a conspicuous postorbital constriction, typical of all apes and most monkeys. Recent man again forms an exception to this rule by retaining his orbits very nearly in their foetal position throughout life and thus acquiring only a minimal post- or supraorbital constriction. In this respect, however, man is not unique, since several

165

platyrrhine monkeys share his absence of any noteworthy onto-genetic migration of the orbits.

Of the endless variety of developmental changes those of the dentition have been studied perhaps most intensively, not only because they can be readily and clearly examined, but also because they serve generally for the estimation of the physiological age of living primates as well as of skulls. The first deciduous teeth to erupt are invariably the middle incisors which appear at or very soon after birth in most monkeys, at an average age of about three months in the great apes and even later in man. The lateral incisors and first molars are the next milk teeth to be added in all primates studied so far, and these are followed by the canines and finally the second molars in monkeys, gibbons and man. In the great apes this final sequence has become reversed, since the canines are usually the last of the deciduous dentition. The order of eruption of the per-manent teeth has evidently undergone considerable changes among primates through a general trend to accelerate the replacement of milk teeth relative to the addition of molars. In some of the rapidly maturing primates the dentition still becomes first of all enlarged by the eruption of *all* permanent molars before any of the deciduous teeth are replaced. In a variety of prosimians and a few monkeys the eruption of the last molars has become delayed until part or even all of the milk teeth have given place to permanent ones. In the great majority of monkeys and in all apes even the second molars do not appear before the eruption of the permanent incisors, and the milk molars become exchanged for premolars before the permanent canines and last molars break through the gums as the final stage of their comparatively slow dental development. In recent man, finally, the extreme of the same trend has been reached inasmuch as all, or at least nearly all of the milk teeth have been replaced when the second molars are being added, and even the first permanent molars are quite often preceded by the eruption of the permanent central incisors. For man, the most slowly maturing primate, it is of special advantage that his small and delicate deciduous dentition is being replaced as soon as possible, while the addition of the grinding molars can easily be postponed. With the gradual lengthening of the period of dental eruption and of longevity among primates, the milk teeth as well as the misnamed permanent teeth have come to serve for correspondingly longer age spans without any improvement in their durability, as is evident from the very common and extensive breakdown of the dental apparatus in old apes and, particularly, man.

The main developmental changes of the skin and hair follow the same general pattern in all primates for which relevant information has become available, but the relative ages at which particular features appear or disappear can differ widely in some of the major groups or sometimes even according to species. The first indication of skin colour can be recognised on the head and back early in foetal life in monkeys, decidedly later in apes and not until the very last stage of prenatal development in the coloured races of man. Post-natally there occurs a noticeable and often long continued increase in pigmentation in species with at least partially coloured skin. In many different primates, however, the skin never does become any more pigmented than in northern Europeans and in a variety of other species certain areas of the body surface remain 'white' throughout life, particularly the eyelids, lips and anal zone. In really dark-skinned primate species, like some chimpanzees, even the palms and soles can become deeply pigmented and the colour can extend in irregular distribution onto the gums and the palate. The varied instances of localised special pigmentation, such as the deep-red 'sex skin' of many macaque species, the sharply circum-scribed red zone on the chest of geladas and the blue nasal ridges of mandrills, do not reach their full colouration before puberty. Among dermal specialisations of taxonomic significance the horny thickenings of the skin over the ischial tuberosities are of outstanding interest because they distinguish all Old World monkeys in contrast to those of the New World. These ischial callosities develop sur-prisingly early in foetal life before any trace of hair has appeared on these sites, even though they have been interpreted simply as an acquired adaptation for sitting. In the Hylobatidae, which without exception share the possession of these callosities, the entire perineal region develops first a coat of fine hairs which becomes replaced by the pair of hairless callosities only at about the time of birth or even slightly later. In the great apes more or less extensive and thick, horny callosities form over the ischial tuberosities only in the course of postnatal life, when they naturally also take the place of a primary coat of hair. Only in man, in whom the thick gluteal musculature has become interposed between skin and bone, have these typically catarrhine features never been recorded for any pre- or postnatal stages.

The first and still unpigmented hairs appear on the eyebrows, to be followed in rapid and regular order by hair on the upper lip and chin and by eyelashes as well as traces of scalp hair. Not until after a definite and considerable interval does the addition of hair on the shoulders, the back and then on the outside of the arms and, finally, that of the legs occur, with the ventral side of the trunk remaining bare the longest (figure 58 and plate 13). As a rule this developmental sequence of foetal hair, called lanugo, begins very much earlier and advances more rapidly in many prosimians and all monkeys than in the great apes, the former being born quite well covered with fairly dense hair, whereas the latter are nearly bare, except for their long scalp hair and comparatively short, sparse hair on the shoulders and back. In newborn orang-utans the hair is particularly scanty and even on the scalp often less conspicuous than in many human babies. Postnatally the hairs generally increase in number, while they decrease in average density per unit of surface. The extreme length of hair, which has become a specialised feature on the scalp of some species of tamarins, macaques and at least some human races, on the back and limbs of orang-utans, male hamadryas baboons and geladas, and on the back and tail of certain guerezas, never reaches its full development until puberty or even later. Such excessive regional hair growth is evidently controlled by sex hormones, just like moustaches and beards which, incidentally, are not limited to man. An extremely old male hamadryas baboon at the Zürich zoo gradually lost every trace of its formerly glorious mantle, so that its coat of hair became almost indistinguishable from that of females.

The general hair colour of many primate species can be radically different in infants from that of their parents. During adult and particularly old life the colour of the hair coat can also change very markedly. In gorillas and chimpanzees, for example, the general appearance of the hair often becomes quite light in old individuals, partly through an increased number of white hairs replacing black ones, but also through loss of pigment in most of the hair. This age change is so marked on the back of old male gorillas that they are commonly referred to as 'silver-backs'.

The interesting fact that man has retained hair chiefly on the scalp, eyebrows, borders of the eyelids, lips and chin, or precisely on those places where hair first appears in all primate foetuses, has often been quoted in support of the so-called foetalisation theory of the late Dutch anatomist Bolk. According to this theory, first advanced in 1926, most of the profound and comprehensive specialisations of man can be traced to developmental retardations and the consequent retention of foetal characters into adult life. The central position of the cranio-vertebral joint, characteristic of adult man, is another classic example of evidence favouring Bolk's views. The latter had the merit of having pointed out that ontogenetic retardations can be very effective evolutionary factors, but today we know that these have operated not only among the progenitors of man, but have played equally significant roles in changing other species in regard to different features. Furthermore, frequently one can demonstrate ontogenetic accelerations in some characters, occurring side by side with retardations in other respects. Many examples of this fact have already been referred to. In man, for instance, the extremely early fusion of the central and navicular wrist bones, the comparatively early coalescence of sternebrae, and the unusually early prenatal descent of the testes are only a few of the many ontogenetic accelerations which have developed alongside the various marked human retardations. To put it briefly, it can be claimed that any modification in the complicated programme of individual development, characterising a given primate group, represents a perfectly valid phylogenetic specialisation and this regardless of whether or not it leads to a change in the adult conditions of the species. For instance, the alterations in the order of eruption of the teeth, discussed above, leave the composition of the final dentition unaffected, yet they are clearly specialisations typical for definite species or groups. On the other hand, any change in the rate of growth of the phalanges of the toes is an ontogenetic specialisation resulting finally in the greatly lengthened toes of orang-utans or the degenerate toes of man.

As a general rule the later phases of the changes with age seem to become modified more readily than the earlier ones which are retained more tenaciously even in entire groups of related species or genera. Among the early developmental changes there persist many which seem to serve no vital purpose as preparation for any later function, but merely represent the conservative repetition of ancestral *ontogenetic* conditions. Such occurrences are often restricted to brief transient stages of development, but can also remain throughout life in vestigial form in all or in only occasional individuals as so-called atavisms. Various examples of such vestigial or atavistic structures have already been referred to (third eyelid, carpal hillock, etc.) as inherited partial or arrested developmental

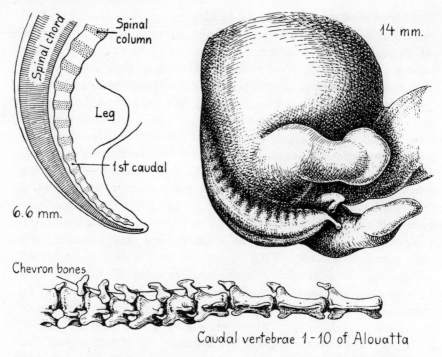

Figure 66 Diagrammatic sagittal section through caudal end of a human embryo of 6.6 mm crown-rump length and sketch of caudal region of an older human embryo of 14 mm crown-rump length, showing external tail with terminal filament. Below: first ten caudal vertebrae of an adult howler monkey, showing the ventral chevron bones.

conditions of ancestral forms, serving no useful purpose. This fascinating repetition of ontogenetic processes during evolution deserves to be illustrated once more by the very appropriate story of the human tail. As shown by figure 66, man still develops an outer tail early in his embryonic existence, which reaches far below the beginning bud of the leg and contains more segments than later persist. In human embryos, only a few millimetres long, eight or nine caudal vertebrae are laid down in this genuine outer tail, but in adult man these have become reduced mostly to only four, forming the so-called coccyx, which is completely hidden in the surrounding tissues. Incidentally, this ontogenetic and phylogenetic reduction has gone to even greater extremes in gibbons and the great apes, which average only two to three caudal vertebrae in later stages of growth. During the first few weeks of embryonic life the outer tail of all 'tailless' primates grows more slowly than the trunk, but still remains clearly visible. Human embryos of about 14 mm crown-rump length bear at the tips of their tails a slender caudal filament, just as in long-tailed monkeys during their prenatal development. It is most likely this filament which is occasionally retained at birth in human babies, though very rarely a much larger tail can persist and contain even striated muscles, capable of moving the appendage. On the ventral side of the proximal caudal vertebrae of all primates there regularly appear so-called haemal arches or chevron bones which straddle and thereby protect the main caudal artery of the tailed monkeys (figure 66). In man the same elements still develop during embryonic life, only to become resorbed again during the earlier part of foetal development. In at least five cases, reported so far, such chevron bones had persisted in adult men as clear atavistic, utterly useless structures. It is not to be doubted that precisely similar conditions as in man will be found in the 'tailless' apes, once their embryonic development and the occurrence of vestigial caudal remnants in later life can be studied with as adequate material as has been available for man. This prediction is based on the confident assumption that all apes and man must have had tailed ancestors at some stage of their evolution. They may have lost their tails at different times or tempos, but in this problem we lack as yet any evidence from fossil remains. Comparative ontogenetic data may become helpful in this connection, as is suggested by the following observation: At the place of the early prenatal disappearance of the tail from the body surface of all recent hominoids there remains a small and normally short-lived prominence, which does occasionally persist to later stages of development as a button-like outgrowth directly over the tip of the deeper coccyx. Such coccygeal tubercles of variable size are found very rarely in older foetuses or young infants of man, but had persisted with much greater frequency among the many bodies of chimpanzees, examined by the writer, including even several juvenile specimens.

In this chapter on growth and development it remains to be mentioned that deviations from the regular, normal conditions are by no means limited to civilised man, as has been widely assumed, but are also surprisingly common in nonhuman primates in their native habitat. It seems particularly necessary to call attention to this fact, because faulty development had commonly been regarded as resulting chiefly from mutations which become rapidly eliminated again under natural conditions. The present intense interest in the causation of human malformations is due to the realisation that normal prenatal development can be profoundly interfered with by various physical and chemical actions, notably radiation, drugs and some virus infections. It is of special interest, therefore, to point out that wild monkeys and apes are also subject to an equally wide variety of tragic deviations from normal development and that not infrequently they manage to survive with even severe congenital defects. This conclusion was forced upon the writer in the course of his collecting activities in the field and was supported by his review of the relevant reports in the literature (published in 1956), from which a few examples may be briefly quoted here. The total assortment of malformations in nonhuman primates is probably just as extensive as that encountered in man, of whom far larger series have been examined. As in the latter, so in the former, albinism is quite common, harelips have been found, undescended testicles (= cryptorchism) occur in different species and so do grossly malformed urinary organs, all sorts of congenital hernias have been recorded as well as instances of double-headed and of cyclopian monsters, and so forth. Digital malformations, ranging from supernumerary fingers and toes to fused, incompletely formed and entirely lacking digits, exist in astonishingly high frequency in a great variety of monkeys and apes. In a series of 113 wild gibbons shot from one locality in northern Thailand there were no less than eighteen specimens (= sixteen per cent) with at least one, and often several, marked digital abnormalities and many additional instances have been found in another large collection of wild gibbons. Some of these cases, which correspond closely to digital malformations reported for man with far smaller relative frequency, are shown in figure 67. In one gibbon of the first-mentioned series the left arm had remained undeveloped, except for a very small part at the shoulder, a condition known in the teratology of man as unilateral peromelia. This fully grown ape had been observed to climb with dexterity and was well nourished in spite of its severe handicap. Some years ago the writer received the body of a young chimpanzee from West Africa which had fallen out of a tree during the pursuit of its group by native hunters. It was distinguished by an abnormally high and short head and the lack of both thumbs and both first toes. Upon dissection it was found that the frontal and parietal bones were not separated by coronal sutures in consequence of a distur-

bance in embryonic development which had also interfered with the
normal separation of the early hand- and foot-plates into digits. The

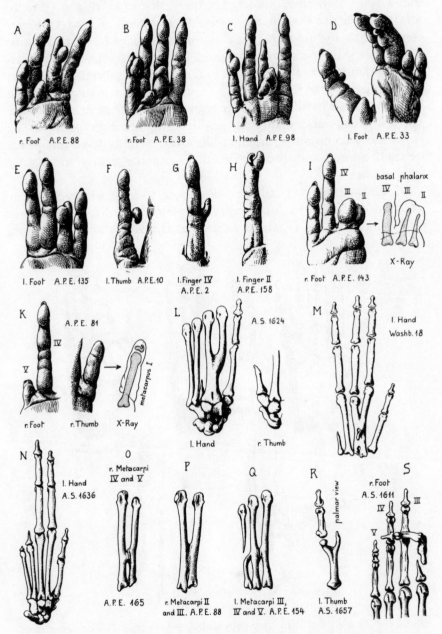

Figure 67   Digital malformations in wild gibbons (after Schultz, 1944).

same syndrome of so-called acrocephalo-oligodactylism is well known to occur in man. The skeleton of a young wild mandrill in the writer's collection shows a grossly distorted thorax, due to severe malformations in the early development of the vertebral column, ribs and sternum, yet this animal managed to survive until it was shot. Minor and more localised abnormalities in the skeletons and the dentitions of wild monkeys and apes have been recorded in astonishingly large numbers, which indicate that their percentage frequencies can hardly be smaller than in man at least in some of the species. Congenital malformations in the dentition seem to be particularly common in the man-like apes, ranging from impacted, twisted, stunted and displaced teeth to supernumerary and totally lacking ones.

Of special interest are all those deviations from usual development which are classed among congenital malformations, if occurring rarely, but must be regarded as normal characters as soon as they

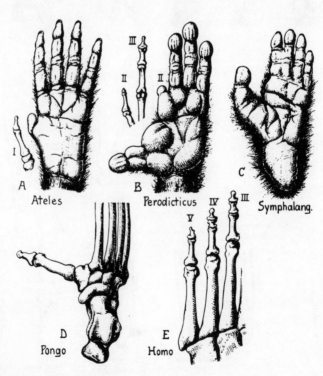

Figure 68   Examples of digital variations, abnormal in most primates, but more or less 'normal' in the species illustrated (after Schultz, 1956).

appear in the majority of cases in a particular species for which they form the statistical norm. A few examples of this sort are shown in figure 68, all appertaining to digits. Normally spider monkeys have eliminated the outer thumbs, but in twenty per cent of *Ateles belzebuth* and in eight per cent of *Ateles geoffroyi* there still exists a perfectly useless rudiment of this digit, as in the specimen *A* of the figure. In pottos it is the second finger which has become reduced as much as in specimen *B* in the great majority of the cases and in the related *Arctocebus* the outer index finger has as a rule even completely disappeared, conditions which in man would be regarded as clear malformations, but in these prosimians they have become normal species characters, still varying individually to a considerable extent. The foot of the siamang (*C*) shows the cutaneous webbing between the toes II and III as a local retention of an embryonic condition which has become normal in the ape, while in man it is found rarely as a harmless hereditary anomaly. The occurrence of only one phalanx in the first toe of the orang-utan (*D*) has been found among these apes in roughly sixty per cent of the cases, so that this developmental defect represents the statistical norm in this species. The analogous congenital lack of a middle phalanx in the fifth toe of man (*E*) has been reported as existing in very considerable percentages of cases, ranging up to eighty in some human races, and forms another example of initially abnormal development changing into normal conditions within a race, species or even genus through increasing frequency of its occurrence.

Chapter 12

Reproduction, Morbidity and Mortality

THE MAIN functions of the organs serving for the propagation of the species are naturally the same in all primates, but in the detailed construction of these organs there have nevertheless developed many different specialisations which reveal nature's remarkable versatility in modifying homologous structures and this often without gaining any decisive advantages. It would not fit into the scope of this book to include here an anatomical description of the male and female internal sexual organs in the various major groups of primates, but the external organs will be briefly discussed as interesting examples of significant taxonomic characters, often having undergone diverging evolutionary specialisation.

In contrast to the conditions in many other mammals, the unerected penis of primates is not attached to the abdominal wall, but is a so-called pendulous one, and the testes are not retained in the abdominal cavity, but migrate ontogenetically into an external scrotal sac. The latter lies at the sides or even in front of the penis in various prosimians and all marmosets as well as in some species of gibbons, whereas behind the penis in all other primates. The testes descend into the scrotum at widely varying relative ages, though always before sexual maturity. In some prosimians, such as *Ptilocercus, Perodicticus* and *Loris,* there seems to occur an intermittent, seasonal descent and in some young monkeys the testes can readily be withdrawn again from the scrotum to just below the inguinal opening through which they had migrated from the abdomen. As a general rule the permanent descent occurs earlier in the higher than in the lower primates, being earliest in man, in whom it normally takes place late in foetal life. The size of the testes shows remarkably little relation to the size of the animal. The percentage relation between the weights of the testes and of the entire body is twice larger in howler monkeys than in spider monkeys, about ten times larger in macaques than in langurs and proboscis monkeys and nearly six times larger in a mandrill than in orang-utans according to the writer's records for a great many fully adult male simian primates. In chimpanzees the testes weigh as much as 250 grammes, whereas in a wild shot gorilla of many times higher body weight only 36 grammes. Owing to the diminutive testicular size of gorillas their scrotum forms merely an inconspicuous swelling, hidden in the fur, in striking contrast to the large pendulous scrotal sacs of chimpanzees and many other primates. From recent studies on macaques it appears that the relative size of the testes may not be constant during the mature life of some monkeys, since at least in the rhesus there occurs a seasonal cycle not only in testicular size, but also in spermatogenesis, which attain their full development only during the height of the annual breeding season of these monkeys. However, even during the periodically minimal size of the testes of these macaques they still are far larger than in, for instance, langurs at any time. Closely corresponding seasonal changes in testicular size have also been found in several prosimians from Madagascar, especially in *Microcebus* and *Cheirogaleus* with their marked alterations in sexual activity.

As in most placental mammals, the penis of nearly all primates contains in its distal part a so-called baculum or os penis, which is lacking only in some treeshrews, tarsiers, woolly monkeys, spider monkeys and man. In most prosimians this bone is comparatively large and may even be forked at its end, while in the great apes it has become reduced to a slender rudiment, measuring, e.g., only 11 mm in a full-grown gorilla, but 23 mm in a mandrill and 21 mm in a potto of the writer's collection. Surprisingly marked species differences in the shape and relative size of the penis bone have been found among macaques and are probably also of taxonomic significance in other groups. The glans penis of many, but not all, prosimians bears small scales and small or large recurved spurs in often complicated patterns, resembling similar developments in various other mammals. Among the Simiae of postembryonic stages only the spider monkeys seem to possess such specialisations on the penis in the form of many small cornified barbs, covering the entire shaft.

The female external genitalia also show many generic and even specific peculiarities which, however, consist mainly in different relative sizes of the various parts and their detailed topographic arrangement. At the ventral margin of the vulval orifice there is always the clitoris, which corresponds to the glans penis of the male.

In many prosimians and some New World monkeys the clitoris contains in its tip a small bony baculum which in some species is even bifurcated, just like the larger os penis of males. The urethral canal completely penetrates the clitoris in a variety of prosimians, thus simulating the male arrangement, but in the great majority of primates it reaches only to the base, as it does in most other mammals. As a rule the clitoris of primates is small and inconspicuous or hidden between prominent labial folds, but in some prosimians and platyrrhines it greatly surpasses the flaccid penis in length and thereby has often led to incorrect sex identifications of living animals. This is particularly notorious in the case of spider monkeys which are distinguished by an extraordinarily long, pendulous clitoris, often contrasting also by its light colour. The pudendal labia are highly variable in their development, especially the folds corresponding to the labia majora of human anatomy, which usually disappear during early growth in many monkeys and in the great apes, but remain well developed in gibbons and siamangs, whose female external genitalia resemble the human vulva more closely than they do the pongid one.

As has already been mentioned briefly in a preceding chapter, the female genital region of many catarrhines undergoes very marked periodic changes in size or colouration. Great enlargements are characteristic of all baboons, mangabies and chimpanzees at definite stages of their menstrual cycles. In many, but not all species of macaques there occurs only moderate swelling, but an extreme reddening of the so-called sex-skin, affecting not only the entire genital region, but also the root of the tail, the upper parts of the thighs and sometimes even the chest and face. Among all the guenons only the small talapoins, often regarded as belonging in a separate genus *Miopithecus*, have acquired this typical, cyclic, oedematous enlargement of the female pudendal structures. One species of guereza (*Colobus verus*) has also been reported as being distinguished among its close relatives by marked sexual swellings. Nothing corresponding to it has ever been observed in any other Colobinae nor in the Hylobatidae. In orang-utans also there occur no cyclic sex swellings, but during the late part of pregnancy such swellings can appear and even reach a very large size. Though evidently correlated to the ovulatory cycles, sex swelling can reappear during pregnancy among chimpanzees and in very old female baboons it can become permanent and attain such grotesque proportions that the poor creatures can no longer sit down.

In the anatomy of the internal sex organs of females the form of the uterus is particularly noteworthy since it shows among primates different stages of fusion in its embryonically double origin. In all prosimians the lower part has become united, while the upper one remains divided into two horns which taper into the right and left oviducts. In the primitive treeshrews these horns are proportionately longer than in other prosimians, particularly the tarsiers. In the Callithricidae there persists a slight indication of a bicornuate uterus in the form of a median groove, but in all other Simiae the uterus has acquired an entirely undivided body and cavity, except for the rare and anomalous cases which have been found in some human females.

The functional changes in the mature female sex organs of primates often show an approximate monthly rhythm. These cycles are accompanied by changes in the uterine lining and in vaginal secretions, as well as by menstrual bleeding in the catarrhines. Very slight and rather irregular cyclic bleeding has also been found in some platyrrhines and in tarsiers living in captivity. Since these cycles are directly related to the maturation and discharge of ova, the possibilities for conception as well as the optimum sexual receptivity of the female are bound to these same periodic changes, generally referred to as oestrus cycles. For instance, the periodic pudendal swelling in chimpanzees, caused by the accumulation of intercellular fluid, marks the height of sexual receptivity of the female. In various prosimian genera a vaginal orifice appears only for a few days during their oestrus periods when it even becomes temporarily surrounded by a turgescent rim, which leaves no trace at other times. These conditions, naturally, limit all reproductive activity to the brief occasions at which the female has become prepared for it.

Actual breeding is still restricted to quite definite seasons in a considerable variety of lower primates, as in so many other mammals. Treeshrews have been found to breed at any season in captivity and this repeatedly in the same year, but in Thailand and Borneo pregnancies seem to occur with at least markedly differing frequencies at different times of the year. Among most of the lemurs of Madagascar copulatory activities must be seasonally limited since births take place during certain months only in many of the species which so far have been adequately investigated and also because the testes undergo marked changes in size, evidently in preparation for a mating season. Furthermore, *Microcebus* and *Cheirogaleus* spend part of the year in a kind of aestivation with all activities greatly reduced and even a remarkable drop in body temperature. According to the latest reports the young of *Microcebus* are all born between November and February in Madagascar, but, when kept captive in Europe, between May and July, indicating the strong influence of environmental factors. For bushbabies it has been reported that they have definite breeding seasons, limited to brief periods, but in captivity births have occurred in practically every month and this in northern as well as southern countries. That female *Galagos* can breed only at intervals is proved by the fact that their vaginal orifice is completely closed for long periods. Ovulation and the accompanying vaginal opening has been found to occur only twice a year in *Galago senegalensis* of the Sudan. Incidentally, such findings require experienced investigators, since prepubertal and anoestrus females resemble males on account of their closed vagina and long peniform clitoris, perforated throughout by the urethra. That wild tarsiers can breed at all seasons has been established on the basis of thorough examinations of extensive collections of preserved specimens.

Among New World monkeys the Callithricidae most likely breed during a limited season only, judging by the results of some field collections, but in captivity they can reproduce at any time and even twice in one year. Wild squirrel monkeys also seem to have a preferential breeding season which can become lost under conditions of captivity. Among howler monkeys, capuchin monkeys and spider monkeys of Central America all stages of intrauterine development as well as newborns and infants have been collected by the writer during a brief period only, proving that these platyrrhines do breed during most, if not all, of the year, though there may be differences in the *relative* frequency of pregnancies.

The breeding activity and, especially, its possible periodicity in Old World monkeys has been extensively studied in quite a number of different species. The many births of Hanuman monkeys from India in a South African zoo have occurred indiscriminately in almost every month and the same is true for wild langurs of the same species, though among the latter there seem to exist slight seasonal fluctuations in the frequency of births in some localities. The African guenons also seem to breed generally throughout the year, but within populations living in areas with marked fluctuations

32 Infant orang-utan.

33 Uakaris (*Cacajao*).

34 'Golden lion marmoset'.

35 White-collared mangabey (*Cercocebus*).

36 (*opposite*) Mother and infant chimpanzee in nest.

37 Titi monkey (*Callicebus remulus*).

38 Tamarins (*T. imperator* above, and *T. mystax* below).

39 Uakari (*Cacajao rubicundus*).

in the available food supply a definite concentration of births at one time of the year does occur and clearly affects the age composition of the troops. In captivity, however, newborn guenons have been reported for all months without any preference for particular seasons, and this in both northern and southern zoos. A great many statistically adequate data regarding the dates of birth of macaques have become available, but the results of the different reports vary widely and seem to be influenced by factors which will require further study. For instance, in India rhesus monkeys are born mostly around the months of March and April and some also toward the end of September. In the flourishing semi-wild colony of the same species in Puerto Rico nearly all births occurred during the last few months of the years 1956 to 1958 according to a detailed study by the primatologist S. Altmann, who also mentions that intensive breeding in form of what he calls 'true' copulations began one year in September and another year in October. Inasmuch as gestation in macaques lasts $5\frac{1}{2}$ months, one or the other of these statements seems to be open to doubt, but one can readily accept the conclusion as regards definite concentrations of mating and of births. In the same colony there were 456 births during the later period of 1960 to 1964 and these births all occurred between January and July with the great majority in the even more limited time of only four months. According to the later reports, matings had been observed from July to January and with an outstanding peak in September and October. In the northern environment of the Japanese macaque a clearly recognisable rutting season exists generally from about New Year to the end of March, when the sexual skins and faces of the monkeys turn to bright scarlet and the female pudendal region swells markedly, particularly in the younger oestrus females. An equally pronounced birth season has been very fully recorded for this species, as usually taking place from about the end of May until the middle of August. Captive-born macaques quickly lose this seasonal limitation of fertile matings under the influence of regular feeding, even temperature, close confinement and unnatural social conditions, so that births do occur throughout the year with little, if any, fluctuations in their distribution.

In the South African chacma baboons, intensively studied by the late psychologist Hall, copulations occur at all seasons, but with some periodic changes in frequency, reaching a peak in the southern spring. In the large populations of baboons in Kenya there is also no clear-cut reproductive seasonality, but births do become most common with the onset of the rainy season. Finally, the mantled baboons of Ethiopia, seem to have definite peaks of mating and of subsequent births, but these are not synchronised in different troops, even though they are close neighbours.

According to all present and still scanty information there are at most only faint indications left of any breeding season among the recent Hominoidea. The psychologist R. Carpenter as well as the writer concluded from their extensive investigations of the behaviour and the development respectively of wild gibbons that these small apes become pregnant and are born in all parts of the year. This, naturally, does not exclude the possibility of some seasonal difference in the relative frequency of reproduction. The same conclusion has been reached by the behaviourist Schaller in regard to the eastern gorilla. For orang-utans and chimpanzees in their natural habitats we still lack relevant data in adequate numbers, but in captivity both these apes certainly do breed at any time of the year. In recent civilised mankind, for which abundant birth statistics are available, it has been demonstrated conclusively that slight seasonal changes in birth rates do exist and that these are not entirely due to cultural or social factors, but must be influenced also by environmental, if not biological, conditions. This has become evident from the observation that the highest birth rates are reached in Europe, for example, in the first half of the year, whereas in countries of the southern hemisphere usually in the second half (Zuckerman, 1957). For unknown reasons this season of maximum numbers of births can even shift in the course of years, as it did in Puerto Rico from around May in the early forties to August in the late fifties (Cowgill, 1964). It is not only possible, but quite probable, that such traces of a reproductive seasonality will be found also in the man-like apes with more ample records. All these findings have here been discussed in considerable detail on account of their bearing on social behaviour, which is generally believed to be profoundly influenced by the existence of breeding seasons, even though they may not be nearly as limited and sharply defined as in many other animals. In the order of primates such a periodicity of reproduction has persisted to the present time in clear form in many, but not all, prosimians and much less sharply limited apparently in some platyrrhines as well as in wild macaques. In the remaining great majority of primate species it seems that more or less clear indications of such seasonality can or may still be recognised in many instances by the fluctuations in the relative frequency of pregnancies and births.

As has already been mentioned, the frequency not merely of 'mounting', but of actual copulation can change very markedly with the season. In many prosimians all mating activities cease completely at certain times of the year and in such monkeys as macaques they become exceptional during long periods at least under natural conditions. Copulations are of course not necessarily equivalent to conceptions, as shown by the fact that captive monkeys and apes have frequently been seen to copulate also during pregnancy and lactation, as well as occasionally during all stages of the female sexual cycle, though conception is possible only during one brief time span. It is at the latter stage, which forms the peak of the oestrus period, when the female monkey or ape usually solicits the attention of the male in various ways and when a really successful copulation takes place. The intensity of sexual activities, as manifested by the frequency of copulations, can differ to a surprising degree even among closely related species, being much higher, for example, in chimpanzees than in gorillas, or in macaques than in langurs. That healthy primates have generally a highly developed sexual drive is evident from the fact that in captivity hybridisation occurs with unexpected frequency whenever the two sexes are represented by differing species or, at times, even genera. In 1954 a list of fourteen printed pages had been published of such hybrids among captive prosimians, Old and New World monkeys and gibbons and this list could be greatly lengthened today, even though few modern zoos care to keep different species together.

The act of copulation is invariably performed by most prosimians and all monkeys with the male mounting the female from the rear and clasping the latter on her back. In many species the feet of the male remain on the ground, but in some other species the male grasps the female by her ankles, thus resting his entire weight on the smaller partner. It is among the Hominoidae that new mating postures have appeared and this in an amazing variety, including the ventral copulatory position, with the partners facing one another, an orientation which occurs also in pottos and perhaps other lorisids, but in only very few mammals, such as bats, some otters and whales. Gibbons and orang-utans copulate sometimes while being suspended by their arms, other times with the females lying on their backs and the males leaning toward or over them. The African apes are even more versatile in this respect, especially when sexual satiation is being reached. Female chimpanzees will often crouch on the ground with their limbs folded under, but again they may rest flat on their backs and particularly the 'pygmy' chimpanzees perform this act mostly in a typically man-like manner. From carefully controlled experiments with infantile primates, raised in isolation, it appears that the copulatory activities of at least macaques and chimpanzees must be learned by the young through imitation of their elders, since they can be absurdly awkward in inexperienced individuals.

The number of young at a birth differs considerably among primates, but single offspring is the rule for all species with prolonged parental care. Among the prosimians there still are numerous genera for which plural offspring is typical or at least quite common. For instance, *Tupaia* and *Microcebus* have regularly two or three young at a birth and the writer has found even four embryos in some *Tupaia chinensis,* while the late anatomist Bluntschli had encountered two embryos in each of the uterine horns of many *Microcebus murinus.* It seems likely, however, that in these extreme cases not all embryos develop to term, since no litters of four newborn have been reported so far for these primitive prosimians. *Cheirogaleus* has usually more than one young at a birth and for some species of *Lemur, Galago* and *Loris* twins have also been recorded and even a few triplets, but single offspring seem to be more common. In the remaining prosimians only single births have become known so far, but the numbers of cases observed are still very limited and a stillborn twin is sometimes eaten by the mother. In the entire suborder of Simiae the marmosets and tamarins are the only ones which produce more than one young at a time as a regular occurrence, singles and triplets being exceptional. In all other monkeys as well as in all the higher primates the primitive condition of plural offspring has normally disappeared, though twinning still recurs occasionally with probably an incidence similar to that in man, in whom it is found generally in slightly more than one out of a hundred births. Among captive chimpanzees at least seven instances of twin births and one of triplets have been recorded so far, indicating what is possibly even an exceptionally high incidence. The twins of marmosets are of the double-ovum type, but single-ovum twins, in which one fertilised egg had become completely divided, do also occur among nonhuman primates, as they do in man.

As in the other mammals, the newborn of all primates must be nursed by the maternal milk at least until their dentition and alimentary organs have become able to make use of other food. The mammary glands of the prosimians differ very considerably in number and location, but always follow the embryonic 'milk lines' from the axillae to the groins, somewhere along which the nipples of other mammals are also located. In all simian primates there are normally only two glands and nipples left, which lie invariably on the chest, a restricted position which is rare among other mammals, in contrast to the milk glands in the inguinal region and those multiple ones strung along the ventral side of the entire trunk. The aye-aye is distinguished among all primates by possessing two inguinal nipples and no others. Many treeshrews, lemurs, *Microcebus*, pottos, bushbabies and tarsiers have two pair of pectoral and one pair of abdominal nipples, but in other specimens of these genera and sometimes even of the same species one can find only two pairs of nipples with either one pectoral or the abdominal pair lacking. The typically simian condition of only one pair of mammary glands, situated on the chest, has been attained by comparatively few prosimians, namely the indris, sifakas and some, but not all, lemurs and loris, according to the observations and reports collected by the writer (1948). It is seen that species with regularly or frequently plural offspring also have more than one pair of milk glands, but so have quite a few other prosimians, notably tarsiers, which produce only one young at a time. The persistence of these superfluous nipples has been regarded as due to a change in function, whereby they may have become merely hafting structures for the newborn to hold on to. This, however, seems very unlikely in view of the fact that the glandular tissue at the base of *all* the nipples has been found to be well developed in some lactating females and that a potto, born and raised in captivity, has been observed to nurse indiscriminatingly at all six of its mother's nipples. In this connection it is of interest also that the Callithricidae of the suborder Simiae have not retained more than one set of milk glands, though they do have twins and even triplets more often than single births. The pair of pectoral glands of the simian primates is located anywhere between the second and fifth pair of ribs, being usually lowest in man, but not until after foetal development since they make their appearance in the second intercostal space in man and apes. In some species of monkeys the nipples lie so close together near the midline that the nursing young are often seen to suck from both nipples simultaneously.

In orang-utans and some New World monkeys the nipples are nearly hidden in the axillae.

The mammary glands, which function only following pregnancy, form no udder- or breast -like protrusions in lower primates since they are spread out fairly flat under the skin. In the higher primates, however, the glandular tissue is decidedly more concentrated and can become so thick after repeated periods of lactation that the glands of multiparous apes resemble more or less sagging breasts of women, though never in such extreme degrees as are found among the latter. The duration of lactation is remarkably long in many primates and can extend well beyond the stage at which the milk dentition has been completed, as it does also in various uncivilised tribes of man. The process of weaning, however, is a very gradual affair and most primate mothers allow their young to sample vegetable food long before they refuse nursing. Since regular lactation usually inhibits the resumption of the female sex cycles, it also postpones a new conception and thereby prolongs the interval between succeeding generations.

Having described the more or less cyclic nature of some reproductive activities in this chapter and the ages of sexual maturity as well as the durations of gestation and of female fertility in the chapter on development, one could attempt to calculate the theoretically possible rates of population growth among primates. Little would be gained by this, however, besides the realisation that the maintenance of the species through reproduction is amply provided for and that at least the rapidly maturing multiparous lower forms might be expected to multiply with such fantastic intensity and speed that they would reach quite impossible population densities in a short time. Actually no 'population explosions' have ever been observed in any primates, except in man in modern times after nature had been profoundly interfered with by man himself. Female treeshrews become sexually mature at the age of $3\frac{1}{2}$ months, can breed repeatedly each year and produce several young every time after a gestation of only $6\frac{1}{2}$ weeks, yet they are far from abundant in their natural habitats which could easily support much greater numbers. In large parts of India, where monkeys are looked upon as sacred and are suffered unmolested in man's fields and even towns, they have never multiplied anywhere near their potential capacity. In various African countries baboons and guenons are constantly or periodically killed in large numbers, partly for food and mainly to protect gardens and plantations. While in some localities they have been nearly exterminated, in others the density of the remaining populations seems to suffer little, being quickly restored through immigration from adjoining forests. Even in the slowly maturing and long nursing man-like apes the local populations *could* increase readily within the life span of one generation, but they seem at best to remain stationary. For instance, gibbons have normally only one offspring every two years and hence about ten during a mother's life-time, but nowhere have they been known to become noticeably more common in their home ranges. In North Borneo, where gibbons are not hunted by the natives and where the dense rain-forest seems to provide ample food throughout the year, one hears but few of their penetrating calls and finds them much more rarely than most of the various Bornean monkeys which are fully as adept at hiding. The large and long ago established colony of rhesus monkeys on an island off Puerto Rico has offered a unique opportunity for studying population growth in monkeys which had originally been quarantined and treated for parasites, which are well fed and free of all predators, and for which a continuous catalogue has been kept. According to the admirable relevant report by Dr. Koford the births in this colony increase the population annually by twenty to twenty-five per cent, but in the same period deaths decrease the population by seven per cent excluding those of infants. The death rate of the latter averages ten per cent in contrast to the rate among animals between two and five years of age, which amounts to only 4·5 per cent, but increases again sharply in the group of older monkeys. The two sexes are generally born in nearly equal numbers and during early life die at similar rates, but after puberty the general male rate of mortality is nine per cent, whereas the female rate only five per cent, so that considerably less than half of all fully mature animals are males. These conditions resemble quite closely the corresponding ones prevailing in human populations. A marked preponderance of females among adults has been found in many different species of monkeys from the Old and New Worlds.

Under natural conditions the numbers of animals populating a given area for the duration of many generations are determined not simply by births and deaths, but are dependent also upon the frequency of accidents and incapacitating injuries, the prevalence of infectious diseases and parasites, the availability of adequate food at all seasons and the composition of the groups according to age and sex. The influence of prenatal and neonatal mortality on the interval between pregnancies is a further significant factor to be considered. Of the hundreds of pregnancies among the large colonies of baboons, macaques and guenons at the Russian research station in Suchumi nine per cent ended in miscarriages and an additional twelve to fifteen per cent were stillborn. In a large American colony of macaques, maintained specially for the study of reproductive physiology, abortions and stillbirths accounted for one out of three pregnancies. Exactly the same pre- and neonatal mortality was found in a total of sixty-six pregnancies among the many healthy chimpanzees, kept at a research station in Florida. There can be no doubt that corresponding conditions prevail also in wild primates and that, therefore, a large part of their reproductive life is wasted and the greatest possible number of offspring is rarely attained. It has already been mentioned in a previous chapter that the act of birth is a difficult and prolonged experience for most monkeys. Not infrequently it happens that perfectly normal full-term foetuses of captive monkeys suffocate during labour and sometimes that the mother dies of exhaustion, or of ruptures of the uterus, torn vulva and subsequent infections. Proper care of a viable newborn apparently has to be learned by at least the apes and some young primiparous mothers evidently lack the experience to aid their firstborn in reaching the nipples and thus let them starve. This may be exceptional in most species of monkeys, but is certainly not a rare tragedy among the great apes in captivity. Several cases of minor malformations of the mouth and oral cavity in newly born monkeys have already been recorded as having interfered with nursing and led to starvation. During the period of postnatal development viral and bacterial diseases and endless varieties of small and large pathogenic parasites are particularly dangerous until a relative immunity can be acquired or selective mortality has left only the most resistant individuals. Malaria, yellow fever, bacillary and amoebic dysenteries, yaws, filariasis, pneumonias and a long list of other diseases, well known in man, have been frequently recorded in nonhuman primates and often in a shockingly high incidence. Tuberculosis, to which all catarrhines seem to be quite susceptible, is apparently rare in platyrrhines. Man's most frequent virus infection – the common cold – can be readily transmitted to the great apes, but not to monkeys. Atherosclerosis, that dreaded malady of civilised man, has recently been found to have a wide distribution among nonhuman primates and to be not at all rare in older individuals of some monkey species. In nature the prevalence of disease seems to vary to a surprising degree with species, locality, season and the age of the individual. For instance, malaria has been found among macaques with geographically widely differing frequency and in Central American monkeys it existed in the great majority of spider monkeys of all ages, but in only few capuchin monkeys from the same localities. Nearly all wild gibbons of northern Thailand were found to be infected with filariid nematodes, while in Borneo such cases were exceptional. The yellow fever virus has a very wide distribution among wild African and American monkeys, from which it has been especially transmitted to human wood-cutters by way of certain high-flying mosquitos, brought down with the tall trees. In epidemic sized outbreaks yellow fever can quickly decimate entire monkey populations, as has happened to the howler monkeys living on an island in the Panama Canal Zone in a protected, but otherwise wild state. By repeated careful censuses of this population it was shown to have become suddenly greatly decreased. Field investigators of the International Health Division of the Rockefeller Foundation reported that in 1947 in one state of Brazil there was a severe epidemic of jungle yellow fever among the howler monkeys and that many hundreds of these animals had died. In another Brazilian state all howler monkeys had become wiped out through an epidemic in 1937 and almost none could be found ten years later, though the local capuchin monkeys had evidently escaped or recovered, since they were abundant at the time of the later study. That wild apes and monkeys can be carriers of such important infectious diseases as yellow fever, malaria, yaws, hepatitis, rabies, etc. represents grave problems for public health. On the other hand, since so many diseases of man can be readily transmitted to apes and monkeys, the latter have already greatly benefited modern medical research, for which they are being used in alarmingly increasing numbers.

The skeletons and teeth of wild nonhuman primates are available in our big museums in such extensive series that the incidence of pathological changes and of repairs to injuries can be established with a fair degree of accuracy. It may be stated at once that these bodily structures become affected by disease and accidents in close correlation with advancing age. Arthritis, sinus infections, dental decay and fractured bones have not appeared with civilisation in modern man, but are just as prevalent in many kinds of other primates, living under natural conditions. For instance, in a survey of 233 skeletons of wild shot gibbons the writer found marked arthritic changes in seventeen per cent of all the specimens, but among the immature ones this percentage was only two, whereas in the series of the oldest animals with worn teeth and obliterated cranial sutures it was fifty-five, including many skeletons with multiple, crippling changes. Among fully adult wild gorillas more or less severe signs of chronic arthritis seem to be also very common, judging by the great

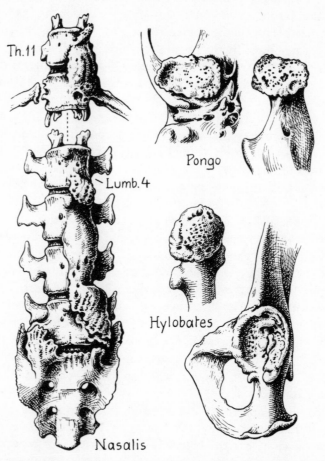

Figure 69   Examples of arthritic changes in skeletons of adult wild primates: large exostoses and fusion between some vertebrae in an old male proboscis monkey, diseased right mandibular joint of a female orang-utan, diseased left hip-joint of a female gibbon (after Schultz, 1956).

number of cases which have been published. For instance, among eight skeletons of old, wild-shot eastern gorillas only one was entirely free of arthritic changes, but several of the others showed them in extreme degrees, affecting chiefly the lumbar vertebrae. Similarly severe vertebral exostoses or arthritic lesions of the hip joints and mandibular joints have also been found in many additional gorillas as well as in adult orang-utans, but in only a few chimpanzees. Chronic arthritis seems to be rare, though not unknown, in most adult wild catarrhine monkeys, except baboons, and has so far not been encountered among platyrrhines or prosimians (figure 69). It is specially noteworthy that all these afflictions appear only with advancing age and seem to progress rapidly with the approach of senility. Still other pathological conditions of the skeleton, well known in man, occur also among nonhuman primates with varying frequency. The inflammatory and distorting disease of osteitis deformans, for instance, has been found in wild gibbons and gorillas as well as in a few monkeys. Sinus infections, leading to pathological changes in the skull, are shockingly frequent among old gorillas and not uncommon in old chimpanzees, being usually associated with alveolary abscesses in the dentition which may have been the primary cause (figure 70). The voluminous maxillary sinuses are as a rule the first to be affected and in the apes often develop openings for drainage through the thin floor of the eye-sockets or into the nasal cavity.

It is exceedingly rare for the 'permanent' dentition of modern man to remain intact into old age, since it has not been the least improved in durability to serve throughout his uniquely lengthened life-span. It is not generally known, however, that the much shorter-lived monkeys and apes are also subject to dental decay which can often become surprisingly prevalent with advancing age and represents a very serious impairment to wild animals, depending on their teeth for cutting and masticating food and occasionally for defence. Alveolar abscesses and carious cavities are almost unknown in the deciduous dentitions of the nonhuman primates, but appear in the second dentition soon after the protection of the hard enamel has been worn away and even nature's repairs with dentine can no longer keep the pulp cavities closed. Chronic irritation of the gums from impacted tough plant fibres produces extensive paradontal disease particularly in the great apes. According to the writer's detailed records on the dentitions of over 2,400 adult and old wild

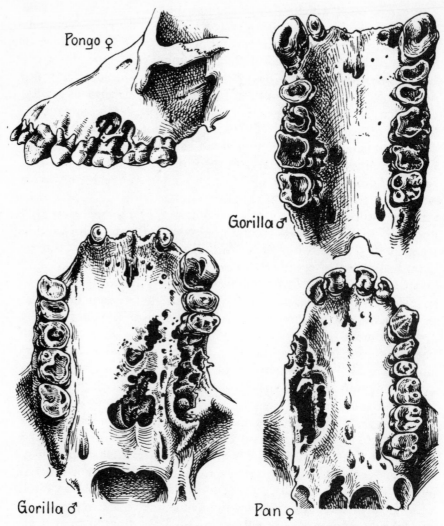

Figure 70 Examples of pathological conditions in the dental apparatus of adult wild apes. The orang-utan has caries and alveolar abscesses on the posterior premolar and the first and second molars; the upper right gorilla has extremely worn teeth with some exposed pulp cavities and alveolar abscesses and one premolar and three incisors lost during life; the lower left gorilla had lost the right canine, both middle incisors and all left molars during life and had developed very extensive alveolar and sinus infections with large drainage openings not only in the palate, but also in the face and left orbit; the chimpanzee had lost most right teeth during life with subsequent sinus infection opening through palate, face, orbital floor and into nasal cavity, there is also caries on all four incisors and an alveolar abscess on the left third molar (after Schultz, 1956).

monkeys and apes alveolar abscesses on at least one and often several teeth exist in the adult, but not old, pongids in anywhere between thirteen and twenty per cent of the specimens, but among the really old individuals these frequencies have risen to sixty per cent. The corresponding percentages for gibbons are seven and thirty-nine respectively. In a new survey of the dentitions of more than 900 monkeys from Central America the writer did not encounter a single case of abscessed teeth in immature specimens and in only four to six per cent of young-adult specimens, but in sixty-seven to seventy-five per cent of the oldest specimens, showing again the extremely rapid increase of these serious afflictions during the terminal stages of life. Caries was found to be comparatively rare in most of these large series of wild primates, particularly among the younger adults, but its prevalence can differ to a surprising degree even in closely related genera. For instance, old gorillas have carious cavities in only three per cent of the specimens, but old chimpanzees in thirty-one per cent and in old howler monkeys caries was found in less than three per cent of all cases in contrast to twenty-nine per cent for old spider monkeys, which subsist on practically the same diet in the same jungles. The complete loss of one or more teeth with subsequent closure of the corresponding alveoli also occurs with far greater frequency in the old-age class than during the prime of life, having been found, for example, in twenty-nine per cent of old chimpanzees and old spider monkeys and in twenty-two per cent of old gibbons. Quite a number of instances have become known of such extensive destruction of nearly the entire dentition in wild monkeys and apes that one is at a loss to explain how they survived under natural conditions in a practically edentulous state and after undoubtedly prolonged suffering. Finally it is mentioned that the large canines of male simian primates are really not very dependable weapons, since they are often extremely worn down and become broken, abscessed or entirely lost with exceptional frequency, even though they are usually the last teeth to erupt completely and, hence, have not served as long as the other teeth.

Anyone who has observed monkeys in tall trees, leaping without hesitancy for breath-taking distances or daring to reach some tempting food at the end of slender, bending twigs, expects that sooner or later they will meet with accidents which tear their hides and fracture their bones. The males of many monkeys have the habit during excitement of violently shaking the branches supporting them – and sometimes these branches will break. Of the Ceylonese gray langurs, thoroughly studied by Miss Ripley, it has been reported that upon the approach of two groups

The males of the troop lead in jumping between branches and trees in a stiff, heavy manner . . . The males jump onto branches they would otherwise avoid – dry, brittle ones that break under their weight. Even at extreme heights they land on these branches, which give way, sending them crashing down to the next level of branches or to the ground . . .

Fortunately even very extensive lacerations of skin and muscle can and do heal with surprising speed and success, often leaving hardly any scars, according to many experiences with captive monkeys. Dislocated joints and fractured bones are repaired or compensated for somehow, if the animal is not killed outright. It is a mystery, however, that wild primates with multiple severe fractures of limb bones, jaws and even the brain case can survive in their jungles sufficiently long for adequate repairs of such seemingly fatal injuries. How many are killed in falls cannot be determined, since corpses quickly disappear on the jungle floor through a variety of causes, but many do survive their agonies, as proved by the numerous relevant descriptions in the literature of repaired fractures in the skeletons of wild monkeys and apes. For instance, in a paper on some adult, wild-shot orang-utans at Cambridge University the late Professor Duckworth had concluded:

That an animal with the jaw fractured on both sides, and practically the whole face separated from the cranial base, should have lived long enough for processes of repair to commence is little short of marvellous. Thus, again, one of the Cambridge specimens has both forearm bones fractured in two places each, with fractured tibia and fibula of the left leg. Yet all these fractures have been naturally repaired.

Among wild gibbons the writer found one specimen with a partly healed fracture of the skull, extending from the foramen magnum diagonally across the cranial vault to the left orbit. In another specimen both hip-bones and one humerus had been completely broken, yet there were well advanced repairs when the animal had been collected. In the entire series of 260 skeletons of wild adult gibbons there are thirty-three per cent with at least one healed fracture. For wild orang-utans the corresponding percentage amounts to thirty-four, but among the far less arboreal African great apes fractures exist not nearly as often, though no more rarely than in many monkeys. In the highly arboreal capuchin monkeys and proboscis monkeys the author found repaired fractures in twenty-eight per cent of the fully adult individuals, but in the much more terrestrial macaques and baboons broken bones are remarkably scarce. Among past and recent races of 'natural man' healed fractures have never been encountered with anywhere near such high frequencies as in many arboreal primates, but, as in the latter, these frequencies increase sharply in old age when bones lose their elasticity and most faculties decline. As a rule, males are more prone to suffer bone-breaking accidents than females, possibly due to their greater weight, recklessness or pugnacity. In many cases fractures heal in a surprisingly serviceable manner and without marked shortening of the bone, but at times limb bones fail to become

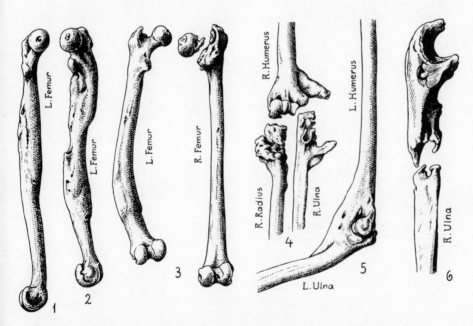

Figure 71 Examples of old fractures in limb bones of adult wild apes; 1 to 5 = gibbons, 6 = gorilla. 1 = fairly well healed diagonal fracture, 2 = double fracture with much shortening, 3 = left femur fractured shaft with bending and shortening, right femur of same individual fractured neck with false joint, 4 = smashed elbow joint with multiple exostoses, 5 = ankylosed ulna following fractured humerus at elbow joint, 6 = false joint at unrepaired fracture (after Schultz, 1956).

adequately repaired, develop false joints or heal with much displacement and loss in overall length (figures 71 and 72).

A great deal more has so far become known in regard to the variety than about the frequency of diseases and injuries in *wild* primates since few bodies have been thoroughly autopsied in the field. From the rapidly accumulating reports on the pathological conditions of *captive* nonhuman primates it appears that the prevalence of harmful parasites and illness must actually be far greater

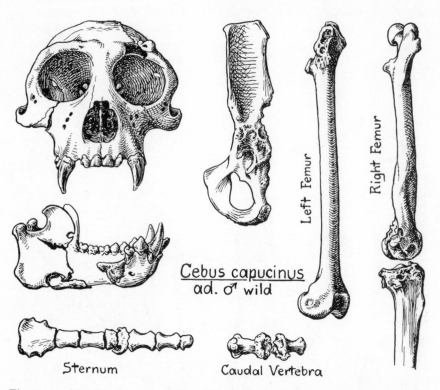

Left Femur

Right Femur

Cebus capucinus
ad. ♂ wild

Sternum

Caudal Vertebra

Figure 72   Multiple fractures and some necrotic conditions of a wild capuchin monkey. The fracture beneath the right orbit extends to the temporal fossa where it had become partly repaired. The frontal sinus was broken open, but the edges have healed, a large part of the right half of the fractured mandible has become resorbed and so has the head and neck of the left femur while the hip-joint has undergone degenerative changes, the right femur is much shortened due to badly repaired fracture and infection of the knee-joint. The sternum, one tail vertebra and two toes also have fractures with much callous formation. This animal was active and apparently well when shot sufficiently long after its accident for all these bony adjustments to have taken place (after Schultz, 1956).

in wild populations than is indicated by the occasional sample studies of their bones, teeth or blood alone. Considering also the manifold malformations, due to faulty early development, mentioned in a previous chapter, it can no longer be doubted that perfect health and normality is far from being such a prevailing privilege of living in a natural state, as has been commonly assumed. If the diseased and 'unfit' can manage to survive, often in such unexpectedly high numbers, there must be many others which succumb to infections or injuries and quickly disappear, unrecorded in any census. The maintenance of a species and the growth of its populations are profoundly influenced not only by the rate of mortality, but also by the incidence and duration of incapacitating misfortunes, which are apt to interfere with, or even prevent, normal sexual activity and thereby reduce fertility. A dominant male chimpanzee with an agonising toothache, an inflamed hip-joint or shaking with malarial chills will hardly respond to even the most alluring sex swelling of his oestrus females. In such highly gregarious animals with lasting close social contacts as the primates, the spread of infectious diseases and the transmission of harmful parasites can be especially effective and rapid.

Mortality due to predation alone would never keep the populations of wild primates from increasing, except for intensive hunting by recent man with his firearms, ingenious traps or long nets. The larger snakes probably do surprise an occasional monkey, but would hardly be left undisturbed to swallow their screaming prey. That snakes are regarded as enemies and carefully avoided by monkeys is evident from the practically universal fear and excitement of the latter in sight of the former. Crocodiles or alligators are undoubtedly a menace to monkeys coming to drink or while swimming. A variety of birds of prey is known to snatch smaller monkeys, if they can be caught in swift flight on an unobstructed perch. Some owls may make similar attempts at night, but this has never been observed. From many reports it appears that the larger cats, particularly leopards and jaguars, will stalk or ambush monkeys and young or isolated apes in daytime and, probably more successfully, at night. In some parts of Africa the persistent killing of leopards by man is supposed to have caused sudden increases in local troops of baboons, but this seems rather doubtful in view of the immediate and very formidable defence of any member of a troop by all the old male baboons, which have actually been known to kill leopards. Even young lions have been seen to chase monkeys on the ground up into trees, from where the latter abuse the former with screams and shaking of branches. Against all these lurking dangers from a great variety of visible enemies, primates can guard themselves by the generally most effective combined alertness and keen powers of observation of their co-operating social groups. It is reasonable to assume that the arboreal species should generally suffer less from predation than do the terrestrial ones. In regard to the terrestrial baboons Professor Hall had summarised his extensive field experiences in his last publication as follows: 'In short, it is doubtful if baboons sustain any but the most occasional loss of individuals to any predator except man. Predation by leopards is probaby greatly exaggerated on the basis of the very rare but dramatic occurrence.' The multitudes of microscopic enemies, represented by viruses, bacteria, protozoa, fungi and countless other parasites, wild primates can only meet with natural immunity or acquired tolerance, besides luck in not encountering them.

The nonhuman primates are all endowed with the *possibility* of multiplying from generation to generation and this even faster than man, because the succession of their generations is more rapid and the period of female fertility comparatively longer. That their populations have generally remained fairly stable, or diminished in recent times, is certainly not due to their being eaten by other kinds of animals, but chiefly because any surplus is being prevented or eliminated by diseases, accidents and crippling injuries, besides the shrinkage of their natural territories through man's need for more land.

It remains to be mentioned here that according to all available, though still scarce, information for nonhuman primates the sexes are born in roughly equal numbers, but in a great many species the males evidently die in proportionately larger numbers to change the sex-ratio of newborn into one of adults which clearly favours the females in varying degrees. A greater mortality of males has been amply demonstrated in most human populations for which accurate censuses are being kept and exists already during intrauterine life, since more male foetuses are being aborted than female ones. In the large colony of semi-wild macaques in Puerto Rico the mortality rate of males exceeds that of females to a significant extent only after puberty and then supposedly due to fighting. Among hundreds of wild howler monkeys and spider monkeys, shot in epidemiological surveys, there were only fifty-seven to sixty-one adult males for every 100 adult females, but among the immature specimens the percentages of males were much higher. A relative scarcity of fully mature males in wild populations of baboons, geladas and patas monkeys seems to be universal and has often been commented on as being due to their slower development, their specialised role in defending the troops with consequently greater mortality, or other more or less plausible causes. Among 247 chimpanzees in various collections the writer found a sex-ratio of 91 males to 100 females in the young and one of only 75 males to 100 females in adults. Wild gibbons are exceptional in having a slight surplus of males which, however, also decreases with advancing age according to the author's data for nearly 600 skulls with known sex and odontologically determined age. The fact that healed fractures have generally been found to be significantly more common in adult males than in adult females of all the skeletal series of monkeys and apes studied by the writer, may serve as an indication that males are more daring and hence have more accidents, including fatal ones. This, however, cannot fully account for the puzzling greater mortality of males, resulting in a preponderance of adult females, even if the so-called 'solitary males' are taken into consideration by the censuses of field studies.

Chapter 13

Sexual and Intraspecific Differences

Secondary Sex Differences

IT IS a well-established fact that postnatal growth as well as the various developmental processes advance somewhat faster and cease sooner in girls than in boys, the differences becoming more marked with increasing chronological age after having been barely recognisable at birth. Corresponding sex differences have also been demonstrated in macaques and chimpanzees, the only nonhuman primates which could so far be studied in adequate numbers and examined periodically after they had been born and raised in captivity. From, as yet, only few reliable records it appears nevertheless certain that in baboons, orang-utans and gorillas also at least the dentition is completed and the growth curve becomes practically level at a considerably earlier age in females than in males, just as in the previously mentioned species. It is most unlikely, however, that such ontogenetic sex differences are characteristic of all primates, because there exist many other species in which both sexes reach nearly the same size in adult life and some in which the females even surpass the males in their average body weight. The accompanying table 6 shows some examples of these enormous species differences in the relation between the average weights of adult females and males. The latter can be more than twice as heavy as the former in baboons, proboscis monkeys, orang-utans and gorillas, whereas in squirrel monkeys, some langurs, guerezas and gibbons this sex difference is exceedingly small and in treeshrews, marmosets and spider monkeys it is even reversed in favour of females. Though it cannot yet be supported with precise data, the writer feels confident that the majority of prosimian species must have practically the same average adult body weight in both sexes, since at least their skeletons are of nearly identical size. It has been suggested that great sex differences in size developed mainly in terrestrial primates

200

Table 6   Percentage relations between the average body weights of fully adult female
males of corresponding primate species according to data collected by the writer in the
and from the literature.

Species	$\male = 100$	Species	$\male = 100$
Tupaia glis	102	Colobus badius	99
Galago crassicaudatus	83	Colobus verus	93
Galago senegalensis	76	Colobus polykomos	84
		Presbytis rubicundus	93
Leontocebus geoffroyi	104	Presbytis cristatus	89
Callithrix jacchus	115	Nasalis larvatus	48
Aotus trivirgatus	88		
Alouatta palliata	81	Hylobates concolor	103
Saimiri örstedii	92	Hylobates hoolock	94
Cebus apella	79	Hylobates lar	93
Cebus capucinus	77	Symphalangus syndactylus	92
Ateles geoffroyi	103	Pongo pygmaeus	49
		Pan troglodytes	89
Macaca mulatta	69	Gorilla gorilla	48
Macaca irus	64	Homo sapiens (ca.)	89
Papio papio	43		

through the selective advantage of large males as effective defenders
of the groups, while arboreal primates supposedly have nothing to
gain from such a sex difference. This hypothesis might appear
reasonable in the case of the largely ground-living baboons, macaques
and gorillas, but is certainly not applicable to the extremely arboreal
proboscis monkeys and orang-utans, nor does it help to explain the
puzzling and very marked differences between the sex-ratios in
body size of such strictly arboreal forms as howler monkeys and
spider monkeys or capuchin monkeys and marmosets.

Secondary sex differences can develop not only in regard to the
absolute size of the entire body, but also in the proportionate size of
bodily parts. Thus the chest circumference and the shoulder breadth
are *relatively* larger in the males of many primates, the differences
being most pronounced in the species with very distinct general
body sizes of the sexes. For instance, in howler monkeys the chest
girth equals on an average 148 per cent of the trunk length in males,
but only 91 per cent in females, and in orang-utans these same
proportions average 203 and 171 respectively. These sex differences easily surpass the corresponding ones found in man. The
proportionate length of the face is also much larger in males than in
females in all those primate species with marked sex differences in
the size of the dentition, particularly the canine teeth. This is

extremely evident from the example of the mandrill, shown in figure 73, which demonstrates furthermore the much closer retention of infantile conditions in the female than in the male. The relative hip breadth and various other, more detailed proportions appertaining to the pelvis are the only measurements which are usually somewhat larger (at least relatively and sometimes even absolutely) in females in consequence of the very essential widening of the birth canal, discussed in another chapter. In all these respects man is a

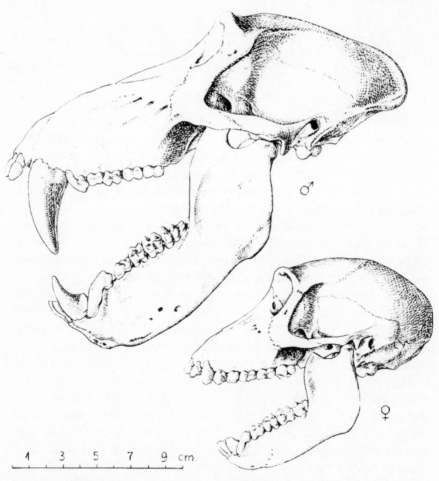

Figure 73   Skulls of an adult male and an adult female mandrill, drawn exactly to same scale.

typical primate in showing the same sex differences in a moderate degree according to his comparatively modest sex differentiation in general body size.

That sex differences in size are closely connected with the intensity of postnatal growth could be demonstrated by the writer for the skull by means of precise cranial measurements on nearly 1,400 primates. It was found that sexual dimorphism is least pronounced in the skulls of such species – as well as in such cranial features – as change least during growth and *vice versa*. For instance, man and various New World monkeys show far smaller sex differences in their skulls than do macaques, orang-utans and gorillas, and the skulls of the former also change with age decidedly less than do those of the latter. Furthermore, the face part of the skull develops generally more marked sex differences than the brain part and it is the face which grows after birth more intensively than the bones enclosing the brain. It appears quite likely that this may turn out to represent a generally valid rule, according to which the more infantile a bodily feature remains the less marked will become its sexual distinction in adults and the smaller the postnatal changes of a species are the less conspicuous remain its sex differences.

In the great majority of adult monkeys and apes the sex can be readily determined by the size of the canine teeth which are most strongly developed in males, as shown by the extreme example of the mandrills in figure 73. In general this secondary sex difference is most marked in catarrhines, among which there are many species with no overlapping in the range of variations of the dimensions of the male and the female intact canines. The best-known exception exists in man, whose canines are of nearly equal small size in both sexes. All gibbons and siamangs are distinguished by having specially long and slender canines with practically the same measurements in females as well as males. Comparatively few New World monkeys have this dental sex difference as pronounced as it is in most of the Old World monkeys and at least in many marmosets the female canines equal the male ones in their moderate size. Finally, among prosimians, the canines rarely seem to form a reliable means for sex determination.

Secondary sex differences in the amount and distribution of hair are by no means limited to human beings, among whom they are generally best developed in Caucasians. In most prosimians the sexes are remarkably similar in regard to their coats of hair, but in

Lemur macaco the hair colour differs strikingly in males and females, being nearly black in the former, but in the latter usually a golden brown. Among the platyrrhine monkeys, which contain many gaudily coloured forms, these features are as a rule most pronounced in males. The latter are also distinguished in certain species by the possession of moustaches, as in some marmosets, or of beards, as in howler monkeys. Fully grown males among catarrhine monkeys, for example, the geladas, hamadryas baboons and some species of macaque develop thick manes of long hair on the head and neck and often also on the shoulders as well as far down the back. These impressive capes serve probably less for protection against the cold than against antagonists. They are poorly developed in females, cover only part of the body, but form an effective armour for the most vital bodily parts of fighting males. Old males gradually lose their mantles and then look mangy and pitiful. There exist a great many other localised and more or less constant sex differences in the patterns, distribution, length and even colour of hair among primates. For instance, in mature male gorillas the entire chest becomes almost bare and with advancing age there appears a striking depigmentation of hair on the back. Beards and moustaches develop in many, though not all, adult male orang-utans, but in only a few females. The males of some species of guenons also have beards even of a very conspicuous colour. The hoolock gibbons are distinguished after puberty by growing a strikingly large preputial tuft of hair in the groin of males, which in some local populations is usually nearly white and thus contrasts sharply with the surrounding dark hair and becomes as conspicuous as the pubic hair of man.

The catarrhines particularly contain many species whose males are much more impressive and decorative than the females not only in regard to size, teeth or hair, but also through various other sex-linked developments. An extreme example of this sort is seen in mandrills in which the males have gaudily coloured skin on the face, buttocks and scrotum, besides the curious large corrugations on both sides of the nose. In the closely related drills these male distinctions have remained more modest without any disadvantage to their very similar mode of life. As implied by their names, the proboscis monkeys have absurdly prominent noses which become many times larger in adult males than females (figure 54). While the noses of the latter change little from their condition in infants, those of the former undergo such a cumbersome overdevelopment that no reasonable explanation for it has been found so far. Equally puzzling in regard to their function are the cheek pads of orang-utans, which are far more frequent and much larger in males than in females (figure 12). In all male foetuses of orang-utans examined by the writer, these pads are already indicated by fine perpendicular skin folds on both cheeks, but these folds do not become more noticeable until the age of puberty, when they usually swell into thick and high plates of connective tissue, flaring out on both sides of the face in variable shapes and not infrequently even meeting one another on the forehead. Another but merely quantitative sex difference of orang-utans exists in their throat pouches which can attain huge sizes in males, extending far into the axillae, while in females they normally remain much smaller. Siamangs possess similar, though less extensive pouches in both sexes, and in these animals they are known to become regularly inflated during their incredibly loud calls.

Among platyrrhines the howler monkeys possess a well-marked quantitative sex difference in their throats, represented by the unique overdevelopment of the entire hyobranchial apparatus, which reaches its extreme in males. It is interesting that this sexual dimorphism has become much more pronounced in the red howler monkeys (*Alouatta seniculus*) than in the other species of the same genus (figure 74). In all howler monkeys the uppermost part of the sternum has become split and forked as a very necessary accommodation for these enormously hypertrophied structures which do help to produce the loud and far-reaching sounds, hardly to be expected from these moderately-sized male monkeys. Even though this is still a very incomplete survey of secondary sex differences in adult primates, one generalisation may be emphasised, namely that by far the most pronounced and diversified morphological differences have developed among the catarrhines, with the degrees of sexual differentiation varying very widely in this group according to genus or even species.

Sex differences in the behaviour of nonhuman primates have been studied intensively in recent years, including many species living under natural conditions. In a great variety of primates it appears that the mature females devote their time chiefly to the care of their offspring while the fully grown males attend to the discipline within and the defence of the group. A division of these fundamental duties is generally quite clear-cut in all such species with

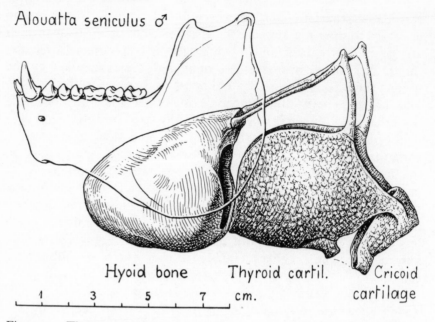

Figure 74   The enormous bony and cartilaginous structures of the throat in an adult male red howler monkey in relation to the mandible of the same specimen, all reduced to the same scale.

marked sex difference in size, strength and dentition as exists in gorillas, baboons, macaques, patas monkeys, some langurs and howler monkeys. Not nearly as marked or consistently followed sex differences in behaviour have been found in species with little or no sexual differentiation in body size, such as gibbons, some species of the Colobinae, spider monkeys, etc. Marmosets and night monkeys are notorious exceptions to the rule of maternal devotion to the young, since in these species it is usually the male who takes charge of the offspring, except of course during nursing, and lets the young ride on its back until the combined weight of the frequent twins or even triplets of marmosets nearly equals the father's own weight. Not only in man, but also in many other primates the males do much more serious fighting than do their mates, as is evident also from the fact that crippled fingers, torn ears and other results of minor injuries have been found with higher frequency among the former. These and further sex differences in behaviour will be dealt with in a following chapter.

Intraspecific Differences

Racial differences occur among primates, as they do in any other natural group of animals, and originate from the same causes, namely as adaptations to environmental conditions through selection, together with reproductive isolation. The determination of more or less marked distinctions between specimens from different localities has been a favourite occupation of primate taxonomists for very many years and had produced steadily growing lists of newly named types. Only recently and with the aid of even larger series of specimens has it become evident that many supposedly new species are in reality merely geographic races or subspecies, characterised by features of limited constancy and significance. Most species with an uninterrupted and extensive distribution tend to differ somewhat in hair colour, average size, relative length of tail or ears, etc. in the peripheral parts of their range, but such differences can often be recognised only in the majority of individuals and then are found to grade into one another among very large series from all localities of the total range of distribution of a particular species. The large island of Borneo, for instance, is inhabited by only one species of gibbon in which several local races can be distinguished by their fairly different hair colour. The latter can hardly have had any selective value in the uniform Bornean jungles, but must owe its local differences to the lack of interbreeding between widely separated populations. Subspecific differences appear not only among species with a very wide distribution, but can often result just as readily from isolation through effective separation by rivers or swamps, mountains, discontinuous forest, etc. In all such isolated groups of a species inbreeding can quickly produce significant changes in the frequency of all sorts of characters with at least potential racial value. Not only such obvious features as hair colour or average body size can become altered in locally isolated populations of wild primates, but also many less easily detected conditions, such as the incidence of the congenital lack of the last molar teeth, the persistence of certain cranial sutures, or the occurrence of many other skeletal peculiarities, of which the technical literature contains innumerable reports.

Species with very wide and at least formerly continuous distribution are still being 'split' into separate species by many taxonomists, whereas 'lumped' into single species with merely subspecies by others according to their varied experience with and evaluation of the relevant differentiating characters. For instance, more than thirty species of chimpanzees had been described in the past, but today most authorities recognise only three subspecies of *Pan troglodytes* and these are almost impossible to define in a manner applicable to all specimens from their supposedly typical home ranges. The night monkeys from Central and South America are often also listed as several different species, or as only one with several geographic subspecies. The very widely distributed baboons of Africa south of the Sahara, which show an unusual adaptability to changing environmental conditions and seem to have respected few barriers in their migrations, do differ markedly in widely separated regions, but the total range of their geographic variations has very few gaps justifying a clear separation into many different species.

That isolation can at times be followed by very rapid morphological changes has been well demonstrated by the example of a guenon, the so-called green monkey, of which a few individuals had been brought from their West African home to the Caribbean island of St Kitts some three hundred years ago. After the original pets had escaped they multiplied till they became pests on the plantations. From careful comparisons between the skulls of ninety-five guenons from St Kitts and of a series from the African parent stock, the British anatomists Zuckerman and Ashton showed that in the former the teeth and many cranial measurements had become appreciably larger and also changed significantly in various other details in a remarkably short time.

In the investigation of intraspecific variability the possible existence of race-mixture is also an important factor to be considered. As already mentioned, many primates in captivity have been known readily to produce hybrids between different subspecies, species and sometimes even genera. From very recent reports it appears that not only racial, but interspecific crosses between different monkeys with overlapping territories can also occur in nature, though probably only on rare occasions. For instance, the behaviourist Bernstein has described unmistakable hybridisation between the large, short-tailed *Macaca nemestrina* and the smaller, long-tailed *Macaca irus* in Malaya, and another highly reliable field worker, H. Kummer, has published convincing evidence for the occasional interbreeding of *Papio hamadryas* and *Papio anubis* in Ethiopia, Subspecific crosses under natural conditions are more difficult to recognise in the results, but there no longer seem to be good reasons for doubting their possible occurrences.

If primatologists are confronted with so many problems in determining and classifying the geographic races of monkeys and apes, it is not surprising that there exists so little agreement among physical anthropologists in regard to the definition of human races. Man has by far the widest distribution of all primates, the least hindrances to migration and interbreeding and is known to have changed morphologically in various respects within comparatively short periods of time. It is to be expected, therefore, that geographically isolated ancient populations of the single species *Homo sapiens* had acquired differing racial characters, which could have become less distinct again here or there in consequence of later migrations and decreasing reproductive isolation. It is commonly believed that the numerous human races together with the results of their hybridisation, so widespread in modern mankind, have provided a total intraspecific range of variations in excess of such ranges in any other primate species. This view, however, is far too crude a generalisation and unsupported by actual evidence. Some of the interesting results of recent work bearing on the intraspecific variability of primates in general are best presented in the form of the following random examples.

Besides the sexual and racial differences occurring within a species, there also of course exist innumerable individual differences which may be hardly noticeable in some species, but very common and marked in others. To be more precise, it is really the variability of a given character which can differ greatly in its intensity among species. Thus, the body weight or the hair colour, the patterns of the cutaneous ridges on palms and soles or the number of palatine ridges, the number of caudal vertebrae or that of the young at a birth, etc. can vary within very wide limits in some species, while in others these same features are found to be remarkably uniform. It may at once be stated that in general all the man-like apes tend to be decidedly more variable in more of their morphological characters than any monkeys and even man. There are exceptions to this rule, but these are limited to one or a few particular features only. The variability of the human hair form, ranging from straight to curly and spiral, represents one of these unusual features. The very irregular distribution of differently coloured hair in some local populations of spider monkeys is another example of this sort.

209

Exceptional variability in generally quite uniform species of monkeys seems to be particularly apt to develop in features which are limited to a given species and hence must have become specialised rather rapidly without having been fully stabilised. For instance, in the peculiar snub-nosed monkey *Rhinopithecus* the bones at the base of the turned-up outer nose differ in their precise arrangement to a most unusual degree, with hardly any two skulls being alike in the nasal region, as is shown by the examples in the accompanying figure 75. On the other hand, such basic conditions as the numbers of vertebrae in the various regions of the spinal column are intra-specifically far less variable in man as well as in most monkeys than in any of the man-like apes. The latter are distinguished also by a remarkable lack of uniformity in regard to general body size and build and to the detailed proportions of the outer body and of the skeletal parts. The photographs of the three adult male chimpanzee

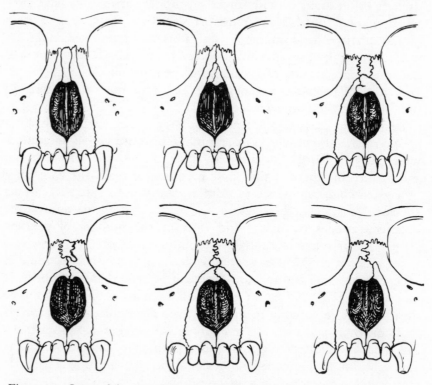

Figure 75   Some of the exceptionally marked and common variations in the bones surrounding the nasal apertures of adult snub-nosed monkeys from one local population (after Schultz, 1957).

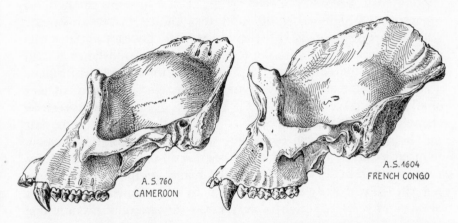

A.S.760
CAMEROON

A.S.1604
FRENCH CONGO

Figure 76   Striking differences in the shape of the skull in adult male western gorillas (after Schultz, 1950).

skulls from West Africa in plate 12 show so many marked individual differences that formerly one would have been tempted to regard them as representing separate races, or even species, but today we have learned to expect such variations within most local populations of great apes. Even more impressive are the two cranial variations among adult male gorillas from West Africa in figure 76, which could easily be bridged by intermediate specimens. The breadth of the brain case, expressed as a percentage of its length, which is the widely discussed 'cranial index' of anthropologists, varies individually in a large series of fully grown orang-utans between seventy and ninety-two with a perfectly normal distribution of all intermediate variations. The size of the brain cavity of the skull in adult male western gorillas ranges from 412 to 752 cm^3 (table 4). Variations in body build, elaborately classified in man as differing constitutional types, are as fully marked among wild apes, and can here only be indicated by the examples of an 'asthenic or linear' orang-utan alongside of a 'pyknic or lateral' one in plate 14.

All these intraspecific variations form the primary source for potential evolutionary changes. Any variation, which provides an adaptive advantage in itself or by being genetically linked with some other one of that kind, will become more frequent through selection until such a more or less exceptional variation has changed into one representing the statistical norm by being typical for a majority of individuals within a later population. Characters which have become

211

specialised in one race, species or genus are very apt to show an exceptionally high variability and to indicate thereby that the particular innovation is still incomplete and has not yet reached the same stage in an entire population. This is clearly shown by the highly variable conditions of the phalanges in those digits which have undergone far-reaching phylogenetic reductions in certain species of all major classes of primates, as has been discussed in a previous chapter. Another good example may be recalled in the finding that the number of caudal segments, which has become extremely reduced in all the tailless Hominoidea, is still unusually variable by ranging between zero and six, with as many as nine segments temporarily existing early in development. In sharp contrast to this, the number of cervical vertebrae very rarely deviates and by only one segment from the ancient and conservatively persisting number of seven. A trend toward reduction of the last molar teeth in some platyrrhines and in man can be recognised in the much greater variability of the size and the numbers of cusps and roots of these than of the other molars which have been far less affected by evolutionary change. Modern statistical investigations on the variations in the formation of dermatoglyphic patterns, the ramification of blood vessels, the relative sizes of functionally related muscle groups, etc. have provided much further proof for this contrast between the comparative stability of unspecialised structures and the generally high variability of features representing new acquisitions of a species or genus.

Chapter 14

Behaviour

THE SYSTEMATIC investigation of the behaviour of primates in
relation to their environment and, particularly, to each other is a
very recent and rapidly growing new science, replacing the long
preceding period which had contributed little beyond popular
descriptions of random observations and anecdotal reports by
explorers or hunters. The modern relevant literature of the last two
or three decades already fills a good-sized library, even though it has
as yet dealt with only a very limited selection of primate species.
Enough has been learned to realise that many aspects of behaviour
can vary to a surprising extent not only from species to species, but
at times even among populations of one species living in different
environments or under changed conditions of population density. It
is still impossible, therefore, to give here a generally valid account of
anywhere near all the many-sided and intricate behavioural characters of such highly developed mammals as are the primates precisely
in regard to their social interrelations and adaptations to their
varied habitats.

Individual Behaviour

A distinction between individual and social behaviour is here being
made merely to bring some order into the many different topics to be
discussed and is not meant literally, nor can it be applied consistently. Much of what is really 'individual behaviour' has already
been dealt with briefly in previous chapters, particularly in connection with such matters as posture and locomotion, reproduction,
functions of sense organs or factors influencing distribution.

All primates rest for at least half their lives by sleeping during
the hours of darkness of the long tropical nights in the case of
diurnal species or during daylight in the case of nocturnal or crepus-
cular ones and frequently in addition by dozing for a considerable

time during the heat of midday. The intense physical energy expended by typical wild primates while fully awake is thus compensated for by more hours of relaxation than are normally required by civilised man. The nocturnal primates sleep very soundly and act in quite a dazed way if suddenly woken up in daylight. Pottos, for example, sleep rolled into a furry ball with the head tucked tightly between the feet. Most diurnal species are more easily aroused from sleep, for which they lie down, stretched on their backs or sides, or sit up according to the nature of the support, the available space, and probably also the temperature. During sleep primates could not avoid the risks of predators nor the dangers of falling without careful selection of their resting sites. It is not surprising, therefore, that all arboreal primates are very particular in choosing tall trees with suitable branches high up for safe retreats, to which they habitually resort. Such terrestrial species as baboons, geladas and some macaques, living in open or nearly treeless regions, congregate at dusk on carefully selected ledges, niches or even in shallow caves on high rocky cliffs, inaccessible to their most feared enemy, the leopard, which stalks its prey in the dark. At such places the monkeys assemble for the night, often in very large numbers, crowded together for warmth or for lack of more space. Patas monkeys are quite exceptional in dispersing before dark, when each adult selects a separate tree or bush for solitary, silent safety. The peculiarly specialised, nocturnal aye-aye builds itself a quite elaborate, globular nest from twigs it cuts and carries to a fork high up in some tall tree. These nests are completely roofed over, have a small side-entrance and a floor lined with shredded leaves, and provide room for only one adult. Though the aye-aye spends a lot of work on this remarkable construction, it rarely seems to make use of the same one for more than a few days in succession, according to the recent observations by J. and A. Petter. Most, but not all, of the other nocturnal prosimians, as well as the night monkey, choose hollow trees or clusters of nest-like foliage as safe hiding places during daytime. The great apes, finally, build themselves crude nests, or really platforms, which they construct in a very short time every evening by first bending larger twigs together, held down by squatting on them, and then adding smaller ones, torn from within easy reach. The tops of fairly slender trees seem to be preferred for these nests, probably because they would shake and bend from the weight of any predator and thus alarm even a soundly sleeping ape.

Each animal occupies a nest of its own, except for mothers with young infants, and usually new nests are made daily to benefit from fresh leaves and avoid ants, scorpions, snakes, etc. which might have invaded an old bed. Fully grown, powerful gorillas usually disdain the safety of tree-nests and merely assemble some foliage at the base of a tree for comfortable beds, off the humid ground.

At sunrise the day's activity begins leisurely and normally with defecating and urinating at the edge of the nest by chimpanzees and in the nest itself by gorillas, which have exceptionally dry dung. Arboreal primates have no problems about disposing of their waste products, except under the unnatural conditions of captivity, when it becomes very evident how much they hate to get their skin and hair soiled and how slowly they learn to avoid this. The first thing in the morning monkeys leave their often very chilly sleeping quarters to ascend to higher places on trees or rocks and there to bask in the early sun as well as to locate their fellow-beings nearby. Many species, notably howler monkeys, guerezas and gibbons, will then burst into prolonged, loud calls to notify their neighbours of their whereabouts or to announce a claim to their own territory. These resounding morning songs are most impressive and un-doubtedly help stray individuals to join their companions. As soon as hunger makes itself felt, wild primates begin the day's work of finding, gathering and 'preparing' whatever constitutes their natural and often highly varied food from the trees or on and even in the ground. Since at least the simian species get their needed calories chiefly from vegetable matter, they must spend a large part of the day in collecting an adequate amount of food, which they generally do in quite a leisurely manner.

The **diet** of nearly all prosimians consists in the first place of insects, including their larvae, of spiders, birds' eggs, small lizards and still other, rarer animal food, but few species despise fruit in season, young shoots, certain bark, leaves, and the soft pith of some plants. Tarsiers in captivity greedily pounce upon and tear to pieces living grasshoppers, newborn mice or nestling birds not much smaller than themselves. Monkeys are in general mainly herbivorous, though the smaller platyrrhines, particularly the marmosets, still rely extensively on a diet of insects to supplement the far more abundant plant food. Wild capuchin monkeys also devote much time in eagerly searching for all kinds of bugs by breaking off the bark of dead branches and even tearing open hollow decaying trees. Largely terrestrial species of Old World monkeys never miss a chance to catch grasshoppers and locusts and will systematically turn over large stones in search of ants, centipedes or scorpions. In addition they will dig with their strong nails for edible roots in the ground or gather the small seeds of grasses with deft fingers. Among macaques one species is popularly called the 'crab-eating monkey', because it and some other species living near water have acquired the habit of gathering all sorts of small aquatic animals in shallow pools or creeks. South African baboons also patrol beaches regularly in search of anything edible left by the receding tide. These special tastes will, of course, not keep most monkeys from raiding the natives' gardens for bananas, beans or corn, nor from seeking trees full of wild fruit, or gathering berries and mushrooms. The Colobinae are the most exclusively herbivorous primates, able to thrive chiefly on leaves owing to their highly specialised digestive organs. All apes also live mainly on fruit and young shoots, leaves, etc. Even the unsocial orang-utans will congregate in huge durian trees when the big, spiny and evil smelling fruit is ripe and chimpanzees are attracted in large numbers to wild fig trees in season. Of some chimpanzee populations it has been observed that on occasion they will suddenly catch and eat young monkeys or small antelopes and that they frequently relish ants and termites. Baboons are also known to kill and devour small mammals, though hardly as a regular habit. Water is of course an essential item on the menu of all primates, but a great many of them seem to obtain adequate amounts from watery food, dew or rain collected in odd places of their arboreal homes. Some species, however, are forced to resort to regular pools or streams toward evening, which they do with great caution on account of the danger of lurking crocodiles, pythons or other swift predators. From field observations and intentional experiments it appears that even the higher primates will not readily try to eat any food new to them. Even before nursing infants are weaned they will snatch samples of solid food from the hands or mouth of their indulgent mother and in this way gradually get to know the entire list of edible items used by their species. In captivity the young will generally accept a strange diet more quickly than do most adults, if none of the latter prevent them. It may also be mentioned, finally, that feeding nonhuman primates in modern research laboratories merely on prepared and standardised 'monkey chow' twice daily might bring the highest efficiency, but not eliminate all dietary deficiency, nor will it replace the stimulating exercise of body and mind, from which wild primates benefit all day long while collecting their varied menus under natural conditions.

Most primates find and recognise their food mainly by sight. Odours of ripe fruit will attract them occasionally and nocturnal species must often be guided to animal prey by their sense of hearing. There can be no doubt that memory also plays an essential role in locating and timing the more abundant sources of food, such as wild fruit trees in season or human plantations, recalled from previous successful experiences from within familiar territory. It is characteristic of primates that they obtain their food with their hand and bring it to their mouths with their hands. Plant food is generally plucked by hand, roots are dug from the earth with the hands and any sort of animal food is caught and held fast with one or both hands. To free the hands for these purposes the animals usually sit down or are at best supported by only three limbs, if not, as in some American monkeys, by the prehensile tail alone. Food is not always consumed immediately where it has been gathered, but may be carried to a better resting place in one or both hands, or stored in the cheek pouches of the cercopithecine monkeys. When the shell of nuts or of other fruit has to be removed with the teeth, or if stems of wild celery are to be shredded, they are held by the hands as essential aids in food preparation. It has been repeatedly observed that some clever monkeys will even crack hard shells by pounding them with a rock. Quite exceptional is the ability of the aye-aye to open coconuts with its rodent-like incisors. It does not pluck the nut, but merely tears the husk off at one place and then gnaws a hole large enough to drink the juice and insert its long middle finger to remove as much of the pulp as possible. Many monkeys and apes clean food that has become soiled with earth or sand by vigorous rubbing between the hands and by washing, if water is 'handy'. According to reliable reports captive as well as wild capuchin monkeys, baboons and chimpanzees can and do make occasional use of branches, sticks or other objects to obtain food beyond reach of their unaided hands. Of particular significance is the detailed account by Jane Goodall of wild chimpanzees not only using grass stalks and small twigs for inserting into nests of ants and termites to 'fish' them out, but to bring a supply of such tools from some distance for this special purpose and even intentionally to select and prepare tools of suitable dimensions and with leaves removed.

While many primates will drink directly by stooping down to the water, others will scoop up the water in their hands or, not infrequently, immerse their hairy hands repeatedly into a pool to suck the water from the soaked hair. A baboon at the Zürich zoo showed remarkable resourcefulness in satisfying its thirst from the water in the moat, below the reach of its arms, by turning around and lowering itself until its tail could be dipped into water, which it sucked dry after it had pulled itself up again. This the animal did on many occasions, as the writer himself witnessed. A wild chimpanzee was seen by Miss Goodall to use a bunch of leaves, rolled into a wad like a sponge, with which to soak up the rainwater left in a narrow treehole, and to squeeze it out into its extended lower lip time and time again. All such examples of behavioural originality in nonhuman primates make it easier to understand how early hominids can have managed to adapt themselves to changed modes of life.

It is quite evident that feeding and drinking must occupy a very large part of the daily activities of primates in their natural habitats and also that it is distributed over many hours instead of being concentrated in a few meals, as in the case of civilised man. The amount of food which the mainly vegetarian primates consume is astonishing and causes their body weights to become much greater in the evening than at the beginning of the day. During the hottest hours around noon most diurnal primates rest in some shady place, especially since they have a much lower heat tolerance than is commonly assumed. During these midday pauses in feeding and wandering the young usually indulge in play and the older individuals in lazy body-care, known as grooming.

Most prosimians, whose hands are not well suited for delicate manipulations, clean their fur simply by combing it with their claws and numerous other species preferably clean with their lower front teeth which have been well adapted for that purpose, being long, slender and projecting forward in a parallel row. Monkeys and apes use their versatile fingers for performing this important function by systematically parting the hair to search primarily for possible ticks, leeches or the very rare fleas and lice as well as for removing any thorns or briars. This grooming is evidently a very pleasant sensation, judging by the great amount of time and devoted concentration given to it and by the expression of blissful relaxation on the faces of the animals involved in this process. Self-grooming, naturally, cannot easily cover the entire body, especially since it has to be controlled by close inspection. Mutual grooming, therefore, is also indulged in so frequently that it has become a prominent social and even ceremonial function, closely connected with social relations, as will be discussed in a later part of this chapter.

The outstanding and many-sided role of the hand in the life of primates is shown also by their usual mode of fighting. As a general rule monkeys and apes grasp their antagonists with their hands to pull them to their mouths for quick biting and may even push them away again before releasing their teeth. It is with the hand that a simian mother aids her painful delivery of a newborn by pulling on the emerging infant and then she holds the helpless creature while carefully inspecting and cleaning it. Last but not least, the hands of primates are used for frequent and meaningful gesturing for social communication.

Individual behaviour in the face of suspected or real danger can take many different forms, of which hiding and fleeing are the most common ones for the young and females, and defensive or threatening attitudes for adult males. Upon the discovery of a potential enemy most monkeys will immediately become immobile, thereby trusting not to be noticed while they can scrutinise the intentions of their antagonist. If evidently still unnoticed a monkey may quickly hide behind or above the nearest thick branch, just as squirrels do. If the enemy stares at or approaches the monkey, it will withdraw to what it feels is a safe distance and this is a slow dignified manner, or precipitately, depending on the action of the threatening creature as well as on the nearness of a secure retreat. The precise reactions are influenced by age, sex and social rank of the animal in danger.

Few enemies succeed in surprising an entire group of primates during daylight while at least some individuals are up in trees, searching for food, selecting routes for the easiest climbing, keeping an eye on their companions and, particularly, on the constant look-out for any change in their surroundings. Every experienced hunter has learned how quickly monkeys and apes discover a resting or prowling large cat and announce this fact with typical, ringing calls of warning and abuse. This acuteness of observation, together with the memory for visual impressions, are amasingly well developed, especially in the simian primates. Any healthy monkey will spot its accustomed zoo-keeper when still far away among a dense crowd of strange visitors and many do show their joy and excitement by loud calls. Wild monkeys readily learn to distinguish between harmless women and armed men approaching them, as shown by their very different reactions. The constant and keen visual exploration of their environment brings monkeys naturally to investigate any new object with irresistible interest, but widely varying degrees of caution. Baited box-traps, for instance, are soon approached with the proverbial simian curiosity and are usually soon entered by the most greedy members of the troop.

An excellent memory for observed impressions gives primates their equally remarkable power of orientation by which they can follow time and time again exactly the same routes in the densest jungles and never fail to find their desired sleeping quarters. This same endowment, together with their intense curiosity, enables them to learn by imitation. This is of especially vital importance during growth, when infants watch and imitate their mothers and juveniles all of their fellow-beings in regard to every aspect of individual and social behaviour. The lack of any opportunity to do this will produce gross deficiencies in later social adjustment, especially of mothers toward their offspring and between sexual partners. This has been convincingly demonstrated by means of experiments with captive-born macaques by Professor Harlow and his collaborators at the primate research institute of the University of Wisconsin. The same investigators have also shown that the typically simian infant has an inborn craving to cling to the maternal body and that without the latter it will adopt instead any artificial soft substitute for its close ventral contact, particularly when frightened. This is merely an example of the profound influence of age on individual behaviour, which changes continually with advancing development according to regular time-tables, which have as yet been studied in only a few species. Already, however, one might venture the prediction that the general kind and sequence of at least the early and most marked behavioural age changes will turn out to be quite similar in all simian primates, including man, and that there are merely the great chronological differences, correlated with the different duration and rates of development.

Communication

The interrelations between animals of like kind are most effectively served by communication. Since nearly all primates are decidedly social mammals, mutually understood communicatory signals play a very prominent part in their daily life and accordingly have evolved into intricate systems in all the higher species. Factual information as well as intentions can be transmitted to other individuals of a species in many different ways and may be given by numerous bodily parts to be received by one or another of the senses. The exact meaning of such signals, i.e., their correct interpretation, must be learned early in life because the various species have often acquired considerable modifications in their codes of communication.

That the sense of **smell** has not lost its importance for intraspecific communication in most prosimians, particularly the nocturnal ones, as well as for many platyrrhines, is evident from their frequent possession of specialised scent glands used for identification purposes in marking territories and routes or for locating sexual partners. These glands, situated on either the throat, chest, arms, external genitalia or anal region, the animals press and rub against much frequented branches as regular signposts and so that the odour of the secretion will be detected by others for recognition of mates or as warning of territorial possession. Such olfactory messages compensate for the restriction of vision in the dark or in dense forest, but require periodically repeated applications to be effective, even if merely at close range. It is noteworthy that such marking by scent forms the only delayed communication which a remote signaller can give while not within sight. The special scent glands of the Callithricidae and of various other New World monkeys serve for the same purposes or only for sexual attraction, whenever they are restricted to one sex. Numerous prosimians and even some monkeys also use their urine with which to moisten their paws and thus to mark their pathways for olfactory recognition. The sexual receptivity of mature females becomes advertised by the changed odour of their vaginal secretions and urine during oestrus. This is particularly essential in the great many species which lack the visual signs of this stage in the ovulatory cycle, which become so conspicuous in female macaques, baboons, mangabeys and chimpanzees. That the peculiar odour does convey definite information is evident from the way males sniff at the genitalia of females while the latter are willing to back up to them. Further information through olfaction is obtained also in identifying food and in locating companions in dense vegetation as well as in the relationship between mother and young. Peculiar odours from the special scent glands and from general perspiration of many primate species are often quite perceptible even to human observers at some distance.

The functioning of the **tactile** sense in communicatory actions is difficult to analyse, but there can be no doubt that seeking, avoiding or suffering bodily contacts between individuals have definite meanings in social interrelations. Since the sense of touch has become most highly developed in the hands of primates it is not surprising that mutual grooming with the fingers is a prominent example of tactile communication. It is commonly initiated by poking at the part to be groomed and evidently results in a very pleasing sensation for the recipient besides often conferring a certain social preference or distinction. In some situations the mere touching of the body of one individual by another seems to act as a signal for attention. Among the simian primates the time spent in mutual and purely manual grooming varies widely with the species, age and sex as well as with social rank. For instance, capuchin monkeys groom much more than do spider monkeys, and baboons, macaques and chimpanzees far more than gorillas, while old males are generally more often the recipients than the performers of grooming unlike the females.

A frightened infant monkey loses all fear upon close ventral contact with the maternal body; indeed, even alarmed grown-ups are still comforted by crowding into a huddle, while sexual partners are stimulated for copulation by sensations of touch. Tactile signals, generated by hitting or biting on the part of a dominant animal, usually produces submission without doing much bodily harm. All communications related to the sense of touch depend naturally on the close proximity of individuals in contrast to visual and auditory communications which can serve at long range.

The highly perfected optical system of all primates enables them to rely very extensively upon **visual** communication for their social life. All visible parts of the body can contribute singly or in combination to the great variety of possible optical signals. Even the mere silhouette against the light often permits reliable identifications in experienced eyes. From the general appearance of an individual in size, colouration, etc. its age and sex are recognised and the changes in posture or the mode of locomotion inform about its activity as far as it is visible. With the exact position or direction of the head, limbs and tail a monkey can express its temper or intentions to all its companions capable of interpreting such signals. The elaborate tantrum-like dances of chimpanzees and gorillas occur not only for the relief of tension, but often seem to be deliberately intended as a challenge to some rival or to impress the rest of the group, being accompanied also by the noise of stamping, hitting and yelling. Rigidity of the body as well as rapid movements can transmit definite information and so does the erection of hair or of the penis. The latter condition is not at all limited to sex behaviour but can also express general excitement and in at least some platyrrhine monkeys penal erection represents a ceremonial display, used in establishing and confirming social dominance. Last but not least, in the highly organised social life of many monkeys each individual's exact location relative to others indicates its social rank and hence belongs also to the list of mutual information, possible to optical perception.

Most expressive for nearby signalling are the great many detailed modifications possible in the appearance of the face, especially in the regions of the mouth and eyes, which are never hidden in dense or long hair. The evolutionary development of the well-named mimetic musculature of the face to far greater refinement in primates than in any other mammals has produced unique possibilities for expressing actions, intentions and emotions by visual means. In all non-mammalian vertebrates the subcutaneous musculature, innervated by branches of the facial nerve, has not advanced beyond the region of the neck, but in mammals it has spread to the head and particularly onto the face. Here it has gained extensive connections with the skin while having split into separately functioning parts and thereby has changed the formerly rigid masks into highly mobile ones. Certain muscle portions become attached to the newly formed outer ears of mammals, others centred around the eyes and still others came to serve exclusively the region of the snout and especially the increasingly movable lips. All of the numerous, more or less distinct, superficial facial muscles of primates have been traced back through comparative-anatomical studies to an undifferentiated muscle matrix of the neck in primitive vertebrates. The intricate, fascinating story of the progressive development and differentiation of these delicate muscles in lower and higher primates, including their marked age changes, has been fully described by the late Swiss anatomist E. Huber in 1931, largely based upon his own work of many years. Here it must suffice to point out that the possibilities of clear and marked facial expression are directly related to the degree of specialisation for independent action in separate parts of the facial musculature and that such differentiation has become perfected in monkeys, as compared with prosimians, and has progressed most in the hominoids, among which it has attained its extreme

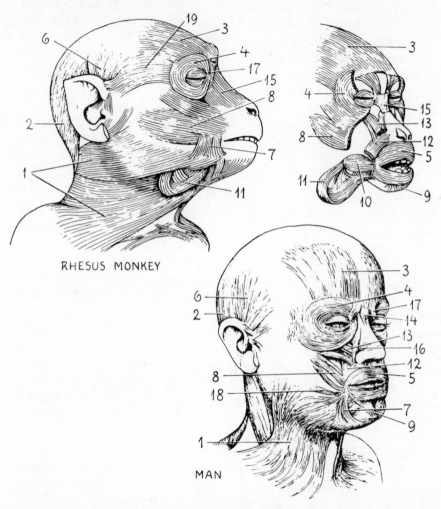

RHESUS MONKEY

MAN

Figure 77　The facial musculature of a macaque in two stages of dissection (after E. Huber, 1933) and of a man. 1 = Platysma, 2 = Occipitalis, 3 = Frontalis, 4 = Orbicularis oculi, 5 = Orbicularis oris, 6 = Auricularis sup. et ant., 7 = Triangularis, 8 = Zygomaticus, 9 = Mentalis, 10 = Buccinator, 11 = Buccal pouch covered by Buccinator, 12 = Caninus, 13 = Quadratus labii sup., 14 = Procerus, 15 = Nasolabialis, 16 = Nasalis, 17 = Depressor supercilii, 18 = Risorius, 19 = Orbito-auricularis.

development in man. This general trend for gradual refinement is particularly marked in the muscles surrounding the mouth and has here culminated in the wide range of expressions and sounds, modulated by the lips, which has been acquired by recent man. On the other hand, some parts of the facial musculature, such as those within and behind the outer ears or that thin sheet which can shift the scalp, have become most reduced in size and function in man (figure 77). Finally, it is also to be mentioned that the range of individual variations in the precise formation of the facial muscles and, hence, in the details of possible facial expressions has clearly increased with the general stage of specialisation and, therefore, has become by far the greatest in human beings.

Among the many static and dynamic signals based upon the face of primates, those of the ocular region play an especially impressive role. Staring with widely opened eyes directly at an opponent forms a central and well understood part of threat behaviour in all the species so far studied, while averting the gaze seems to indicate submission or at least a lack of antagonistic intentions. By raising the brows and lowering the often startlingly light-coloured upper eyelids monkeys such as geladas or mangabeys can give an additional and readily seen threatening signal. This is replaced in some other species by contracting different muscles of the brow region to effect frowning with accentuated wrinkles. Independent of, or accompanied by the just mentioned displays, the precise shape and action of the mouth can greatly increase the repertoire of different facial expressions. With tightly compressed lips, retracted lips, graded opening of the mouth with or without exposure of the teeth, with pouting, lip-smacking and those further modulations of the oral shape which give the effect of grins, monkeys manifest their differing emotions of being pleased, puzzled, frightened, willing to pacify, submit or to defend themselves. Extreme and repeated opening of the mouth, so commonly seen in captive monkeys, is not simply bored yawning, as it appears to human observers, but signifies a warning or impressing gesture.

A redundant assortment of further visual signs, typical for definite situations in primate behaviour, is rapidly becoming known from the intense recent interest in these problems. For instance, spreading of the ears in threatening and flattening them in fleeing or fighting are reflex actions of some, but not all species. Rapid movements of the tongue behind the moderately opened mouth accompany

225

grooming in many monkeys and copulation, just before and during the act, in various others. The direction and curvature of the tail may change its significance with different situations, indicating at times social rank, an invitation for mounting or a readiness for flight, but the precise meaning is probably not alike in all species, partly because the locomotor functions of the tails of so many monkeys overshadow their availability for signalling. Even changes in cutaneous colouration can serve as visual communications for certain catarrhines. The cyclic changes in the colour of the female sex skin of a variety of these primates are of essential help in providing relevant information. The red colour of the chest of male geladas, of the nose of male proboscis monkeys and of the face of males of several species of macaques becomes intensified during excitement, due to nervous dilation of superficial capillaries, as in blushing of the cheeks among human beings. The curious habit of geladas of turning their lips completely inside-out as a warning signal also produces a startling effect with the sudden change of the entire snout from a dark to a shiny light colour. Not only the body itself is made use of for an endless variety of signals, but at times even inanimate objects do serve for communicatory purposes. For instance, the shaking of branches is a widespread habit of monkeys and apes when confronted by antagonists. Branches, broken off while being shaken, are frequently dropped toward, or even thrown deliberately and well-aimed at the enemy below.

In contrast to the great majority of other mammals primates are generally quite noisy in their daily life and certainly do rely extensively on **auditory** signals for communication and not even on vocalisation alone. For example, adult gorillas regularly drum with their open palms on chest or abdomen as loud signals of challenge, and chimpanzees love to produce as much noise as possible by rhythmic stamping, hand-clapping and by beating or kicking a hollow tree in the jungle or an iron door in their cage. Audible grinding of and chattering with the teeth in frustration or anger is quite common in some catarrhine monkeys and has been heard in the exceptionally silent orang-utans as an indication of the same emotions. Greatly alarmed monkeys usually flee without regard to the easily heard rustling caused in the foliage along their rapid retreat, but whenever they believe themselves to be undetected, they can and will vanish with careful silence.

The most frequent sound signals of primates are produced by the vocal organs and associated resonating structures. These *vocalisations* vary widely in kind or type and in the degree of duration and of the distance at which they can be heard. According to recent reports, they include even certain sounds of some prosimians and marmosets which are not perceptible to human ears. The chorus of 'hoots and hollers' produced by fully grown indris, howler monkeys, guerezas, gibbons, siamangs or chimpanzees carries farther than a mile, when sent out, as is often the case, from the tops of tall trees, but most other primates are utterly incapable of anything near such impressive performances. The Lemuriformes are in general a notoriously noisy class of prosimians in contrast to the Tupaiiformes and the Lorisidae. The vocalisations of primates differ as a rule also according to age and sex of the individuals uttering them and, last but not least, the voices of different species of a genus can be very distinct. Evidently, therefore, the wail of an infant or the bark of an old male can be identified at once by all members of a group in regard to its origin and meaning, particularly since they normally seem to acquire an amazingly keen sense for discriminating between even the finest differences in sound. The number of distinct vocal sounds of nonhuman primates has been variously given by different investigators as anywhere between seven and twenty-six, depending on the species and no doubt also on the methods of observing and recording. It seems likely that species adapted to dense forest with very limited visibility have come to rely more on acoustic than on optical communication. It may also be expected that the man-like apes could modulate their voices far better than lower primates on account of the much more advanced differentiation of the oral musculature in the former. So far, however, these assumptions have not been tested by adequate series of detailed comparative data.

The great majority of primates have distinct and far-reaching calls to express, for instance, alarm with barks and distress with screams and many use grunts or growls as signals of warning or, especially, for mutual contact, effective at only a short distance. Such 'contact notes' play a particularly essential and frequent role in the life of the numerous nocturnal prosimians as well as in the only nocturnal monkey *Aotus* which can depend comparatively little on visible signals. Some other close-ranging sounds, referred to as squeals, clicks or grunts, appear to serve mainly to attract the gaze of others, which then can be given more specific visual signs. A shrill single bark is uttered by many monkey species at the moment danger is noticed, but subsequent repeated double or multiple barks signify that the danger has been clearly recognised and may even indicate somehow its nature and direction. Incidentally, it seems that these loud calls in the presence of danger from enemies or competitors are the only ones applied to events outside the group, since all other auditory signals are concerned primarily with intra-group affairs. If it is considered that all such sounds can convey different information, by being repeated slowly or rapidly and given in combination with or without other definite signals, it is realised that monkeys do possess the means for a quite elaborate repertoire of communicative possibilities. This is further increased by the fact that certain vocal signals, no matter how much alike, are differently evaluated, if seen to come from individuals of different age, sex or even social rank. Naturally, all members of a species must be able to imitate their species-specific vocabulary and learn its precise meaning, just as we have to learn the language of our families for social intercourse.

From this very brief survey of the many different means of communication, capable of being combined in countless different ways, it appears that these possibilities are fully adequate for all contingencies of the social life of nonhuman primates. That the unique articulate speech of man with all its sophisticated languages must have evolved from the crude vocal communications of earlier primates is a matter-of-course to most primatologists. When, how and why these ancient beginnings became so highly specialised and elaborated during human evolution are basic and fascinating questions for which we have as yet no generally agreed answers. The recent theory that human speech became possible only with certain topographic changes in the vocal tract, especially the relation between larynx and oral cavity, supposedly connected with the acquisition of the erect posture, is not founded on convincing evidence. The real and main reasons for the development of man's unique ability to communicate with words as symbols are undoubtedly to be found in the phenomenal evolution of his central nervous system.

Social Behaviour

Very few primates lead a solitary life for any length of time and these exceptional species exist mainly among the nocturnal and rather primitive prosimians. The vast majority of primate species live in groups of widely varying size, held together by definite social organisations. It has already been pointed out repeatedly that this coherence of individuals in a permanent group is of great advantage to primates which have comparatively few other defensive means. This social life is continually dependent upon mutual information, given and received through the varied means of communication. For instance, all behaviour appertaining to *reproduction* and especially that leading to mating is directly correlated with visual displays which are often supplemented by olfactory and tactile signals. Not so long ago it was still commonly believed that it must be the continuous attraction between the mature sexes which causes primates to live in closed and permanent contact as promiscuous hordes or in the form of harems. With the intensive recent studies of the social life of primates under natural conditions it has been learned that their mating systems actually differ widely and that many species do have marked seasonal restrictions in their breeding activities with little effect on their social organisation. Lasting attachments between only two individuals of opposite sex, as in gibbons, are as rare as are the extreme harem systems of hamadryas baboons, while general promiscuity, more or less subjected to the priority claims of dominant males, seems to represent the most frequent mating pattern. It has also become clear today that co-operating watchfulness and the protection of the young are of vital importance to most species and are more potent factors in maintaining their social organisation than is sex. As a rule, sexual behaviour occupies only a small period of time in the life of mature female monkeys. Most of them are pregnant about half the year and do not become sexually receptive again until their infants are weaned a year or more after their births. Even while not encumbered with offspring the female is actively interested in sex for only a few days at the height of each oestrus period. The corresponding conditions of the man-like apes are in general very similar, but the duration of gestation and of lactation is more prolonged than in monkeys, so that fruitful sexual performances on the part of the female are even rarer. These generalisations are not invalidated by the occasional occurrence of mating behaviour during or soon after pregnancy, which may be due to exceptional social or abnormal hormonal factors.

The reproductive activity of males also requires very little time, even though the frequency of copulations differs to an astonishing degree among some catarrhine genera. For instance, macaques and baboons copulate far more frequently than do langurs and most guenons and sexual intercourse is much more common in chimpanzees than in orang-utans and in gorillas, yet the rate of reproduction under natural conditions is about the same in all of these species. The amount of sexual activity of males in the course of a year is influenced not only by the seasonal changes in breeding, typical for some species, but even more so by the ratio between receptive females and mature males, which can also differ very widely from species to species and in various groups of a population. Social rank is an additional factor in this connection, since high-ranking males usually claim first place in copulating at least during the brief time of maximum female receptivity, marking ovulation.

In very many, if not most, monkeys and apes it is the female of the species which really initiates sexual activity and solicits the interest of the male for mating behaviour. The relevant feminine wiles may differ somewhat according to species, but they rarely fail to bring results. The female in oestrus will generally back up to the male, briefly crouch down as a gesture of submission and look or even reach back. This may be accompanied by rapid motions of the lips or tongue and preceded or followed by the tactile stimulation of grooming. If not effective at the first trial, these inducements are repeated or applied to some other male. In some species of monkeys, as for example the bonnet macaque, males apparently rarely wait to be invited and act on their own initiative by simply approaching a receptive female, lifting her tail, examining her genitalia and mounting. Copulations may be repeated in quick succession by the same individuals, but are quite commonly followed by further matings for which females as well as males have chosen different partners. While more or less durable consort relationships do occur in some species, they are certainly not the rule in most, as has already been mentioned.

The birth of an infant and its early *parental care* represent the most important events in and for the social life of nonhuman primates. The outstanding behavioural specialisations of at least all monkeys and apes are centred around the successful rearing of their comparatively few descendants. While the mother, naturally, takes care of the nourishment of her growing offspring and usually also of its transportation and most intimate protection and earliest instruction, it is chiefly the male which protects the mother, burdened with a helpless infant, and in exceptional species even carries the latter around or shows other adoptive care of them during their growth. This division of responsibilities, less common in other mammals, forms an effective social bond, even though the mother has generally the heavier duties. The arrival of every newborn arouses intense interest in all members of a typical group of monkeys and throughout its early infancy it continues to be a general attraction, of great help in the integration of the group. Females without infants of their own especially seem to have a strong compulsion to cluster around the new baby in persistent attempts to examine and care for it in spite of the mother's usual refusal. Even subordinate females often seem to rise in social rank or relative dominance over others during advanced pregnancy and, particularly, for some time after they have given birth, when they benefit from the protection of their offspring by the older males.

The proverbial devotion of wild primates to their young is really exemplary in mothers with the accumulated experience gained from having raised previous infants. That this successful care has benefited also from opportunities to observe other mothers of the group, has been demonstrated by the inefficient maternal behaviour, ranging to completely ignoring or even killing of the newborn, of captive mothers raised in total isolation. The urge to hold an infant, even though not their own offspring, seems to be irresistible to many female monkeys, according to the common occurrence of 'baby snatching' and adoption among captive macaques and other species. Indeed, there appeared recently a remarkable report of a wild female spider monkey seen to carry an infant howler monkey for several days until the latter died of starvation.

Immediately after the birth of a monkey-baby the mother, though often quite exhausted from the prolonged ordeal of the delivery, thoroughly inspects the wet newborn and cleans it by licking and grooming, at times even before the afterbirth has been expelled. As a rule and with surprisingly little delay the newborn clings with its strong grasping reflex to the hair on the mother's abdomen. There it may also be supported by the mother with either the hands or the flexed thighs (figure 20 and 78) and soon finds, or is gently led to, the breasts. The close contact between the ventral sides of baby and mother satisfies the former's intense craving for warmth and tactile urges and brings it nearest to the nipples on which it will normally nurse greedily and at first very frequently. Any separation of this contact will at once result in cries of distress of the young

231

Figure 78   Gibbon carrying its infant in its lap while brachiating (after a photograph).

infant. Prosimians with very immature and weak newborn are for some time kept in nests for warmth and protection and, when moved, are carried by the mother with the teeth, instead of the hands which are needed for locomotion. This habit of 'oral child transport', employed during the first few weeks, is typical for tarsiers, bushbabies and *Microcebus* and has been observed even in one

232

species of guereza according to recent reports. It is never long, however, before the infantile prosimians acquire the power to cling to their mothers' fur, at first on the abdomen and later most often on the back. As soon as the infants of simian primates have become sufficiently strong and active they generally move to the mother's back, where the longer and denser hair provides better holds and where they are lesser impediments to the mother's locomotion (figure 22). Infants of such daring brachiators as all gibbons, usually continue to cling to the mother's abdomen while being cradled on the flexed thighs of the latter when hanging by her arms (figure 78). The swift patas monkeys also belong to the minority of species which dislike letting their young ride on the mother's back, though this would seem to be the most secure and least troublesome manner, except for the danger of a rider being swept off while dashing through bushes. Due to the erect posture and the loss of hair on the trunk, the infants of man have to be carried in the flexed arms, freed from locomotion, and later often astride a hip or in a sling or hood on the back.

As the growing infant gradually acquires more independence and ventures farther away from the reach and slowly relaxing attention of its mother, it will still rush back to her at the slightest alarm. Even fair-sized juveniles are rarely refused a ride on the mother's back for a difficult jump or precipitate flight. Though still nursing many times daily, the developing youngster watches its feeding mother with rising interest and sooner or later begins to grab morsels of food out of her hand to munch them himself and thus learn the taste of his future new diet before all his milk teeth have fully erupted. At the highly variable age of weaning, about the time of the mother's regaining her normal sex cycles, any attempts to nurse are actively discouraged and much of the previously so intense maternal care is replaced by renewed interest in sexual behaviour. The youngster, having already learned to find its own food, soon forgets its former dependency on the mother in the close association and lively play with its companions of similar age. In situations of social conflict or other excitement, however, a continued bond between mother and offspring normally persists well beyond the time of final weaning and at night the two remain in close proximity. The fully grown males are the ones which provide the most lasting and effective protection of the immature members of a group long after the mothers have lost their special interest in this responsibility.

233

In other respects the attitude of adult males toward the young of their social units differs very widely among primate species between complete indifference and devoted parental motivation, with antagonistic behaviour being rare before the young approach sexual maturity. For instance, while the proud male patas monkeys will pay no special attention to their offspring, male marmosets regularly carry infants on their backs and sub-adult male hamadryas baboons will steal infants and care for them in the way a mother would handle them.

The urge to congregate in close association with members of the same species is at least as pronounced in the majority of primates as it is among many other mammals. Indeed, the permanency and organisation of *social group formations* seem to have attained unique degrees in a great variety of simian primates in which this represents a necessary adaptation for the protection and survival of these slowly maturing and reproducing species. Among the more rapidly growing and, at least potentially multiplying prosimians there are as yet only moderate signs of the typically simian trends toward extensive social coherence. Particularly in the many nocturnal prosimians one commonly finds only mating partners or mother and young in close association, but more individuals may sometimes collect within signalling distance in the same limited region, if environmental conditions are specially favourable. The diurnal lemurs tend to congregate in somewhat larger numbers and closer contact, rather than in mere family groups. Of *Lemur catta*, for instance, twenty-four and even more have been counted in direct association, of *Lemur macaco* a maximum of fifteen, and of the also diurnal sifakas up to ten individuals. Among monkeys and apes the size and composition of social units vary very extensively, but those consisting of only parents and immature offspring are certainly quite exceptional. One such rare example is not surprisingly the only nocturnal simian species *Aotus,* which is rarely seen in groups larger than single families of two mated adults and one young, with even the parents straying apart at times. *Callicebus* and all Hylobatidae also live in small family groups of parents with their infants and, possibly, younger juveniles, but these are usually not far from other such families. Orang-utans are further examples of exceptionally 'unsocial' higher primates which can manage to survive in their native forests with abundant food and few enemies without the advantages of closer social ties than the sexual and parental ones. In

contrast to these few monkeys and apes lacking pronounced gregariousness, the great majority of the species spend most of their
lives in close contact with anywhere from half a dozen to way over a
hundred companions of their own kind. The highest numbers occur
in general among the moderately to extremely terrestrial forms of
macaques and baboons. For instance, in the open country of East
and South Africa one encounters sometimes more than 100 baboons
in close association, but in forested regions of Uganda the groups of
the same species never attain such a size. The rhesus macaques of
northern Bengal, which are thoroughly at home on the ground, have
been found in bands varying from twenty to over a hundred, and
groups of bonnet macaques range usually from twenty to thirty
monkeys in forested districts, but from forty to sixty outside the
forests. The largest of the langurs, *Presbytis entellus,* which spends
much more time on the ground than any of its many smaller relatives,
lives in bands of up to fifty individuals in forests and up to 120 in the
dry areas of central India. At night the hamadryas baboons of the
Ethiopian highlands have been found to congregate on rocky cliffs
in groups of more than seven hundred individuals, but for their
daily foraging these separate into much smaller groups, consisting
of varying numbers of regular family units. Typical arboreal species
live in fairly permanent bands of anywhere between about ten and
fifty individuals, these numbers being greatly influenced by the kind
of environment, seasonal food supply and competition with neighbouring groups. The average troop of the exclusively arboreal
proboscis monkey numbers about twenty, though at dusk more may
assemble in favourable sleeping trees. Most guenons, guerezas and
other tree-dwelling Old World monkeys live in social bands of very
similar size. For the universally arboreal New World monkeys we
have specially reliable counts of group sizes for howler monkeys and
spider monkeys, the respective data ranging in the former from four
to thirty-five and in the latter from ten to forty. The other species of
the Cebidae seem to form social groups of quite moderate size as a
general rule, but among the Callithricidae some species tend to
congregate in units of considerable size. The eastern gorilla lives in
remarkably stable bands, composed of somewhere between five and
fifteen individuals, whereas chimpanzees have been found in highly
unstable groups of up to eighty individuals, representing the temporary associations of somewhat more steady subgroups of considerably smaller size, according to the latest field studies.

The *composition of groups* according to age and sex can differ among primates nearly as much as does the size of the social bands. Judging by the still very fragmentary data for prosimians, the sexes of adults do not seem to differ significantly in their numerical representation within the mostly small social units. On the other hand, typical groups of monkeys and apes generally contain more mature females than males and the relative total number of young, as well as their proportionate representation at different stages of growth, can fluctuate widely among the species and at times even within the same population, being dependent on possible breeding seasons, prevalence of predators, parasites or infectious diseases and the available food supply. The sex-ratio can change markedly with advancing age, both sexes being usually present in quite similar numbers among the young, whereas in adults there is often a very significant excess of females. This is undoubtedly due to some differential mortality of the sexes, as has already been mentioned and as is well known in the case of man. Not infrequently the age change in the proportion of the sexes within a social unit is also accentuated by the fact that males tend to leave their own groups far more often than do females, turning into more or less solitary, or at least peripheral animals during the ages of approaching maturity or, more rarely, senility. Since the relative number of individuals, which have been temporarily forced out of the groups of their youth by rivals or have drifted apart voluntarily, is usually very modest, it can hardly explain the often very marked sex difference in the numbers of breeding adults. Such stray individuals do occasionally start their own small groups which slowly grow through reproduction upon maturation of adopted youngsters. At other times new groupings can also originate from the sudden splitting of units which have become too large for the territorial food supply or simply for continued normal social organisation.

The general type of social structure of groups of wild simian primates is best shown by the following few concrete examples, based upon detailed reports of field workers, many of which have been collected by De Vore as editor in a recent and most useful book on primate behaviour. In the howler monkeys of Panama the average group contains thirty-eight per cent of infants and juveniles, seventeen per cent of fully adult males and forty-five per cent of adult females, of which about a third carry infants. In Panamanian spider monkeys the young amount to only about thirty per cent of the total group, the adult males to twenty-four per cent and the adult females to the remaining forty-six per cent, but this species seems to have an unusually strong tendency for some adult males to live temporarily in separate unisexual bands. Among rhesus macaques of northern India a typical group is composed of thirty-five per cent of young animals of various ages, twenty-one per cent of full-grown males and forty-four per cent of adult females. For groups of baboons living in open territories of East Africa, fifty-two per cent young, thirteen per cent full-grown males and thirty-five per cent mature females have been given as representative figures. The corresponding data for Indian langurs happen to be very similar when calculated as averages of detailed counts, obtained in different bands, namely fifty-one per cent for infants and juveniles, fifteen per cent for adult males and the remaining thirty-four per cent for adult females. It must be emphasised, however, that the sex-ratio of adults within particular social units of a population can vary to an extraordinary degree. For instance, in wild baboons adult females have been found to outnumber adult males in some groups by only 2:1, but in other troops by 9 or even 10:1. In the howler monkeys of Argentina (*Alouatta caraya*) adults of both sexes exist in nearly equal numbers, in contrast to the great surplus of females among the Central American howler monkeys (*Alouatta palliata*). Finally, the patas monkeys of Uganda normally permit only a single fully-grown male in one band containing up to twelve adult females, plus occasionally one to two sub-adult males. Such sharply organised polygynous one-male family units are also typical of hamadryas baboons and possibly some other catarrhine species. In the fairly stable social groups of eastern gorillas, youngsters of all ages form in general forty-four per cent, adult males, including the one to two old 'silver-backs', nineteen per cent and adult females thirty-seven per cent. For wild chimpanzees, finally, one cannot give comparable data because their group formations are so exceptionally changeable, with adult individuals of both sexes wandering off continually, or being added for varying periods, and with their highly promiscuous mating system. Experienced observers of chimpanzees in the wild seem to agree that in general females outnumber males to some extent among the individuals of breeding age and that all infants and juveniles together can normally form close to half the total population.

Social groups of primates generally live within *territories* of their own with fairly definite boundaries, not transgressed by others at least in the presence of the troop occupying the particular area. The size of these territories can differ very widely for many and chiefly ecological reasons among which food supply, water, shelter against sun or storms and safe retreats for the night are the most decisive ones. Naturally, territorial size depends also on the proximity of neighbouring troops which, in turn, is influenced by population density of the species in a particular region. The part of a total territorial area most frequently and intensively used by a family or troop for its routine activities is its real home range and this in particular is as a rule defended against other groups of the same species by various threatening displays – mostly vocal ones – much more often than by actual fights. As a general rule the territories of species living on open ground with sparse vegetation have to be much larger than those of decidedly arboreal forms, dwelling in dense forests. Among the former there are the terrestrial baboons, geladas and patas monkeys which sometimes wander for many kilometres in their daily foraging. On the other hand, most platyrrhines, guenons, langurs and guerezas rarely range through forest over more than one or two kilometres in horizontal directions, but their intense climbing activities nevertheless demand a great deal of energy. From all accounts available so far, it appears that the social units of the prosimians do live within comparatively small territorial ranges which they mark chiefly by olfactory means. The behaviour of the man-like apes in regard to territory differs as widely as do their patterns of gregariousness. Gibbon families have very limited ranges which frequently overlap, but are fiercely defended whenever intruders meet the owners. Orang-utans apparently tend to range very far and their extremely small groups seem to scatter readily. For gorillas as well as chimpanzees typical home ranges of many square miles have been found, but these are not respected by neighbouring groups and are often not defended by the occupying bands, as they commonly are in monkeys.

The cohesion of primates in definite social groups and their routine behaviour within their territories are regulated by ingrained patterns of carefully observed social interrelationships which differ in many respects from species to species. Even though the complex structures of these relations have recently been explored by intensive studies in many different primates, much remains to be learned before all of the underlying principles are discovered. Already, however, it has been fully recognised that the various individuals in a group can have very different social status, as shown by the *dominant* behaviour of some over others in regard to leadership, priority in feeding and in sex relations as well as in still further actions subject to competition. The social rank of a subordinate individual often manifests itself by the distance it keeps in relation to more dominant animals in its group and quite commonly it is revealed also by certain grooming behaviour. In howler monkeys, baboons, most macaques, gorillas and probably many other species with marked male superiority in size, adult males are decidedly dominant over adult females and the latter are dominant over the young in the group. This, however, is merely a general rule which differs among primates in the precise kinds and degrees of dominance, typical for particular situations, even though the big males invariably do have the most essential social control over their troops. In the almost communal lives of howler monkeys the adult males of a band feed in close association, share receptive females without quarrelling and lead or defend their groups by co-operating action. Within the hierarchial social units of baboons, on the other hand, chasing, biting and fighting occurs continually and seems to be needed for enforcing the strict order of dominance. Gorillas, however, can keep order fully as well by means of mere gestures or grunts, with which the dominant male can usually induce immediate submission of subordinates. The dominance relationships in the unstable bands of chimpanzees are not closely organised in definite patterns, though priority ranking is mostly very clear-cut, being largely determined by individual qualities as they develop with advance in age. Among adult gibbons either the male or the female may be dominant over its partner in controlling the family, perhaps because both are of equal size. The same exceptional conditions have also been found in spider monkeys which do not differ either regarding the size of the sexes.

The relations between different groups of a population of wild primates will rarely degenerate into actual fighting, being mostly settled by means of long-range threatening or mere warning signals, if not by the discrete withdrawal of the smaller party. Territorial intrusions of members of different primate species (except man) are mostly ignored as long as the approach keeps at a respectful distance, which may be only a few metres. From the many recent field studies it has become evident that nonhuman primates under natural conditions and regardless of differences in their social systems, generally lead peaceful lives and are certainly not nearly as aggressive and easily provoked creatures as used to be assumed from their behaviour under crowded conditions of captivity. Accentuated pugnaciousness and abnormal nervous tension have been observed only when a population is intensively harassed or if its density has somehow increased to a point demanding acute competition for food and shelter. These same conditions have commonly been mentioned as excuses for the aggressive behaviour of groups of man.

Chapter 15

Evolutionary Trends

AFTER having dealt in the foregoing chapters with many different aspects of the life of primates of the past and present, it has become possible to review in conclusion some of the most significant evolutionary trends which can be recognised within this mammalian order and, particularly, those trends which have given the surviving species their present conditions. Of greatest interest in such a survey is naturally the biological history of man, as the most highly developed of all primates, a problem which shall here receive special attention.

Probably not less than seventy-five million years ago a group of small and as yet very primitive insectivores must have gained some advantage favourable for adaptation to new conditions of environment through one or another specialisation, no matter how slight. These earliest progenitors of primates possessed most likely a very similar mode of life and type of body build as has been retained with remarkably little alteration by the treeshrews of today. Like the latter, so also the first prosimians must have been of small size with slender trunks, long tails, short limbs, each ending in five digits, the long face with a full set of simple teeth still entirely in front of the brain case, the eyes on the sides of the head and the olfactory apparatus well developed. They were undoubtedly maturing rapidly and relying on multiple offspring for maintaining their kind. From the fossil record it appears that such inconspicuous creatures spread in a short period of time over wide territories of the Old and New World and soon began to become differentiated into an amazingly large variety of forms. While all prosimians have numerous primitive features in common, they do not often represent simply different stages along the same evolutionary trends, but rather the results of nature's lavish experimentation with diverging trends. Of the approximately one hundred valid genera of fossil primates, found so far, over two-thirds belong to the prosimian suborder, a fact which does illustrate their pluripotential endowment and astonishing radiation, even if it is admitted that a larger share of these older primates may have been discovered than of the younger simian ones. The prosimians as a class differ intergenerically to an extraordinary degree not only in regard to body size, mode of locomotion and adaptations to nocturnal life, but also in their manifold specialisations of the dentition, the digits, the regional numbers of vertebrae, the shortening of the face, the position of the orbits and still further significant features, as has been described in previous chapters. Specially noteworthy are such unique innovations as the enormous elongation of certain tarsal bones in the expert leapers, the extreme opposability between the first and fourth digits for powerful gripping in the slow lorisiform types, and the long lists of localised specialisations of aye-ayes and of tarsiers, which have no equivalents among any other primates.

Having evolved into at least seven different families, representing no less than four distinct infraorders, the surviving prosimians have become restricted to regions of temperate and tropical climates in Africa and South Asia after having long ago completely disappeared in America and Europe. Ever since the isolation of Madagascar this huge island had lacked effective predators and competing simian primates until its late invasion by man. Thus it provided an ideal home for the later development of the infraorder of the Lemuriformes with its profusion of strikingly different forms. Similar to the remarkable radiation in the evolution of the marsupials in Australia, Madagascar could freely adapt its early prosimians to every ecological situation and even experiment with such one-sidedly specialised types as the aye-aye or the giant *Megaladapis*. In other parts of the world prosimians must also have been flourishing during earlier stages of their history according to their numerous fossilised remains. Their former abundance and wide distribution diminished greatly in later periods while many of the freely radiating lines of specialisation seem to have become unsuccessful and to have terminated in extinction. The recent prosimians from outside Madagascar have managed to survive only as small, nocturnal and comparatively silent forms, among which the tarsiers are the most specialised types regarding anatomical construction. Judging by various features of the skull and teeth of certain Eocene tarsiiform fossils with notable expansion of the braincase and marked shortening of the muzzle, it seems quite conceivable that the most primitive *simian* primates could have originated from such early prosimians. Whether or not the New World monkeys arose separately from some ancient North American lemuriform type, perhaps similar to the exceptionally complete finds of *Notharctus*, is still being debated largely along zoogeographical arguments.

Just as the prosimians had already developed along so many different lines during the early stages of their evolution, so did the simian primates apparently start their differentiation as soon as they had risen above prosimian predecessors. Even though no undisputed 'missing links' between the two suborders have been discovered so far, it seems most likely that some division in simian characters had already made its appearance early in the Oligocene period and this possibly due to separate ancestral sources among the prosimians. At any rate, the simian fossil remains from the succeeding Miocene period show a clear diversity of forms which must have required a long time to emerge. The few fragmentary fossil finds from South America can all be recognised as typical platyrrhine monkeys, including those of early Miocene age, while the origin of catarrhines can be traced even to Oligocene deposits, according to most authorities.

The long and most likely separate evolution of the New World monkeys has resulted in a great variety of forms, surpassing that of the recent Old World monkeys. Though all of the former are entirely arboreal and at most of moderate size, they have split at an early age into two very distinct families, the Callithricidae and the Cebidae, of which the first is basically quite uniform, whereas the second contains representatives of widely differing degrees and kinds of specialisation. As a group the platyrrhines are distinguished from the generally similar Old World monkeys by having retained three premolar teeth on both sides of each jaw and by not having acquired truly opposable thumbs, besides many other detailed features. They all possess tails of at least moderate length with the great majority having very long tails and several genera even prehensile tails which function much like a fifth limb in locomotion and tactile exploration. The family of marmosets and tamarins is composed of only small, strictly quadrupedal species, capable of climbing by means of their sharp claw-like nails. They have lost their third molars, except for the single species of the somewhat aberrant genus *Callimico*. In general they are decidedly more primitive than the other and larger platyrrhines and usually bear

243

twins or occasionally even triplets. Their brains are less well developed than those of other recent monkeys and accordingly their mental manifestations are comparatively dull.

The surviving representatives of the second family of New World monkeys, the Cebidae, still contain a single nocturnal form in the archaic *Aotus* and another quite primitive genus, *Callicebus,* both of which have not shared fully the advances in brain development, behaviour and other specialisations, characterising the numerous remaining cebids. Some of the latter have acquired new modes of locomotion, particularly the closely related spider monkeys and woolly monkeys which have become highly expert brachiators with a correspondingly great elongation of their upper limbs. In this respect, as also in their remarkably large relative brain size, they correspond quite closely to the Hylobatidae of Asia. The squirrel monkeys and capuchin monkeys resemble in their general body build and some aspects of their mode of life the macaques and guenons of the Old World and thus are further illustrations of the extraordinary parallelism with which the evolution of the Western and Eastern simian primates had proceeded. Besides such general trends, there have occurred many specialisations, limited to one or another platyrrhine group, such as the enormous enlargement of the hyoid apparatus in howler monkeys and the unique perfection of a prehensile tail in the same genus as well as in two others. It is furthermore of interest that, with the exception of howler monkeys, secondary sex differences are much more modest in platyrrhines than they are in a great many catarrhines and that the former have generally a denser coat of hair, even though they are not so well adapted to withstand cold as are the latter.

During the long evolutionary history of catarrhine primates in the immense area of Eurasia and Africa there had occurred many profound terrestrial and climatic changes which provided ample reasons for migrations or distributional separations as well as for new adaptations to altered environmental conditions. According to the direct, but very scant evidence of fossil finds, chiefly of teeth and jaw fragments, it appears that the two catarrhine superfamilies, the cercopithecoid monkeys and the hominoid higher primates, had separate origins, or at any rate, had already become distinct in the Oligocene period. On the other hand, all the indirect, but very comprehensive comparative anatomical evidence of the recent catarrhines could readily be explained on the assumption that the

244

hominoids had evolved from the basal, very early and unspecialised stock of monkeys.

The entire group of Cercopithecoidea is comparatively uniform, though several types have become well adapted to terrestrial life while the great majority have remained chiefly arboreal. The size of adults ranges between that of a few guenons and macaques, smaller than the largest platyrrhines, and the huge male mandrills, equalling chimpanzees in weight, besides some even larger Pleistocene species. All Old World monkeys have only two premolars left on each side of both jaws and have acquired thick horny pads covering the expanded lower ends of their hip bones. They are without exception experts in quadrupedal running, climbing and leaping, can also move below branches by their arms alone, if necessary, and readily stand upright on their hind legs for a higher level of vision. The proportionate length of their extremities has remained moderate, especially that of the arms, and their trunks are slender and flexible owing to the long lumbar region, usually still containing seven of the largest vertebrae. The shoulders, hips, thorax and sternum have never become nearly as broad as in the higher primates and the tail is often very long, but in some species, chiefly among macaques and baboons, it has become much, or even extremely, reduced with apparent ease. Visual acuity has been highly perfected in all Old World monkeys, but their olfactory apparatus is rather poorly developed. In the majority of the species the sexes differ very strikingly in regard to body size and other sex-linked characters, besides the much larger canine teeth of males, typical of all cercopithecoids. As far as is known, these monkeys are all born in a much more advanced state of development than are any of the higher primates, but receive devoted maternal care during their entire infantile growth, Their prenatal and main postnatal periods of life have become significantly lengthened in comparison with these conditions in prosimians, as has happened also in the more progressive platyrrhines.

The strictly vegetarian subfamily of Colobinae, composed of the Asiatic langurs, proboscis monkeys and their rarer snub-nosed cousins, as well as of the African guerezas, is distinguished by the development of an elaborate sacculated stomach. In the other subfamily, the Cercopithecinae, are collected all guenons, mangabeys, baboons, macaques, etc. which are characterised by the possession of cheek pouches and the ability to adapt themselves to a remarkably wide range of environmental factors, including different climates and diets.

The only remaining and the most highly evolved group of all primates, the superfamily of Hominoidea, shares a great many characters with the Old World monkeys, a fact which favours the assumption of one common origin for all catarrhines. For instance, all have exactly the same dental formula and have acquired a fully rotated, opposable thumb and all show equally directed, but differently progressed, trends for the enlargement and perfection of the brain, for the refinement of the mimetic musculature and for the prolongation of the main periods of life with the accompanying long parental protection of the young, to mention only a few random examples. The least highly developed hominoids, the gibbons and siamangs, still possess ischial callosities, like those of all catarrhine monkeys, but these comparatively small structures do not appear until much later in life and equivalent horny pads also develop in many juvenile great apes. It is furthermore significant in this connection that cyclic female sex swelling, typical of a considerable variety of Old World monkeys, occurs also in chimpanzees among the hominoids in a fully comparable form.

As a taxonomic group the hominoids are decidedly more diversified than their nearest relations, the cercopithecoids. The small gibbons are veritable dwarfs alongside the huge gorillas and the extremely dense hair of the former contrasts enormously with the remnants of hair in man. In regard to locomotion the recent hominoids have come to differ very widely, with the Asiatic apes representing the most versatile arboreal acrobats and best brachiators. The small hylobatids, however, move very frequently by brachiation and with reckless rapidity as well as admirable grace, whereas the heavy orang-utans quite rarely and hesitatingly, ordinarily preferring slow quadrupedal climbing. The African apes have remained less obviously adapted to brachiation and are far more at home on the ground where they commonly progress quadrupedally and then necessarily supported on their knuckles. Man, finally, has evolved into a unique erect biped with the arms freed from any locomotor function. The differentiation of the sexes has also developed along diverging lines among the hominoids. Both sexes are of practically equal size in the Hylobatidae and so are even their very large canine teeth. Among the Pongidae the body size of males greatly exceeds that of females in orang-utans and gorillas, but not nearly as much in chimpanzees, and the sizes of their canines differ in a closely corresponding manner. The secondary sex differentiation of man is very moderate in comparison with that of the two largest species of the great apes since the average body size of men surpasses that of women far less and the canines are equally small in both sexes.

Among the recent catarrhines the entire group of the hominoids is clearly distinguished by a very pronounced widening of the trunk, so that the shoulders, thorax and hips have become proportionately much broader than in any monkeys. Consequently the clavicles had to elongate and the iliac blades of the hip bones to broaden out. The sternum has also changed from a universally slender row of separate bones in lower primates to a broad structure whose bony segments become united sooner or later during postnatal life. The shoulder blades have shifted from the sides of the narrow chest of monkeys to the broad back of all apes and man. The pelvic girdle of the hominoids has migrated toward the thorax in consequence of the shortening of the lumbar region with its uniquely reduced number of segments. On the other hand, the average number of only three sacral vertebrae, typical of Old World monkeys, has increased to over four in gibbons, to five in man and to even well above five in the great apes. The vertebral segments of the caudal region have become reduced in all recent hominoids to a few vestiges, completely hidden in the surrounding tissues, except early in prenatal life. After birth the spinal column begins to invade the thoracic cavity in the higher, but not in any lower primates, and thereby approaches the centre of gravity when the trunk is held upright. This ontogenetic process advances most in man. These and various other anatomical innovations of the trunk have followed the same trends in all hominoids and represent adaptations facilitating the erect posture which these primates can readily maintain in sitting, climbing and during occasional or prolonged bipedal locomotion.

The limbs of the hominoids have generally gained a remarkable range of mobility in all directions, as if to compensate for the reduced flexibility of their comparatively stout trunks. In their relation to trunk length the limbs of all recent higher primates have become longer than in the lower catarrhines, particularly the upper limbs of all apes and the lower ones of man. It is very significant that the arms of man have participated in this prevailing trend and that they can even be proportionately longer than in some gorillas, while the human legs have not become nearly as much elongated as have the arms of the Asiatic brachiating apes. The hands and especially the feet of hominoids have undergone various diverging specialisations which become most marked only during postnatal growth. These adaptations for different functions include extreme lengthening of the digits II to V in the gibbons and orang-utans, whereas unique shortening and partial degeneration of the first toe in orang-utans and of the second to fifth toes in man.

The evolution of the hominoid head has been dominated by the enlargement of the brain which has tended to increase the neuro-cranial vault more than the base and has reached an outstanding extreme in recent man. In addition to the increased total size of the brains of apes and man, very marked progress has also taken place in the elaboration of the convolutions of the cortex and of many further neuroanatomical specialisations. The heads of the great apes are furthermore distinguished by the enlargement of the face in accordance with their powerful dental apparatus, most marked in males. The joint between the head and the vertebral column, situated near the centre of the cranial base in foetuses of all simian primates, shifts during later growth toward the rear not only in monkeys, but also in all apes, whereas in man it retains its nearly central, foetal position throughout life, whereby the human head becomes fairly well balanced on top of the erect spinal column. A corresponding lack of ontogenetic change exists also in regard to the position of the orbits of recent man, which remain underneath the brain cavity to adult life, whereas in the apes they migrate forward with the face after infantile growth and then produce striking postorbital constrictions and necessitate protruding supraorbital bony shelves. The skulls of hominoids are characterised also by a much more extensive formation of sinuses than in any catarrhine monkeys. Typical hominoid diversity is evident in the head also by the widely differing development of the noses of gibbons, gorillas and man and furthermore by the relative size of the outer ears, which are extremely small in orang-utans and enormously large in most chimpanzees.

The important trend to prolong the main periods of life, which is clearly recognisable in the entire order of primates, has progressed most in the recent hominoids, though even here in different degrees, corresponding to the general evolutionary advancement of the three families. The periods of pre- and postnatal growth surpass in gibbons those of monkeys and in the great apes those of the gibbons. The gestation period of man has lengthened little beyond that of some pongids, but the duration of his infantile and juvenile periods has reached the maximum degree of prolongation whereby the attainment of sexual maturity has become most retarded in man. In spite of the very significantly increased duration of prenatal life in all hominoids, their state of maturity at birth has not progressed nearly as much as it has in monkeys. The newborn great apes are remarkable also for their small size not only in comparison with the size of the mothers, but also in relation to the exceptionally ample diameters of the pelvic birth canal. The increased life-span of higher primates has not been accompanied by a greater durability of such vital structures as for example the 'permanent' dentition, which tends to break down frequently and extensively in old hominoids. Such senile degenerations have become most common in recent man, whose longevity by far surpasses that of all other primates and extends way beyond the age of female fertility in contrast to the corresponding relation in apes as well as monkeys.

The most intensively studied primate family is, naturally, that of the Hominidae and the most interesting problem that of man's evolution. The facts bearing on the latter and the story reconstructed therefrom have already been described by countless authors, especially in recent years. In concluding the present account of the life of primates in general, the writer feels nevertheless that a condensed summary of the main evolutionary trends having affected man, and of his present status as a primate, is not only justified, but expected. There certainly can no longer be the slightest doubt that man must properly be assigned to the superfamily of the Hominoidea and must have evolved from the same common ancestral stock which also gave rise to the recent man-like apes. When, where and why the hominid line had diverged from the lines leading to other extinct or surviving higher primates is still a much debated problem. The relevant conclusions have to take into consideration not only the direct, though very incomplete, evidence of all fossil remains in their not always certain chronological sequence, but also the wealth of phylogenetic deductions from comparative anatomical and embryological investigations as well as the most recent accumulation of data on chromosomes, blood, parasites, etc., in addition to the known geographic changes of the past, which had influenced primate distribution. It is not surprising, therefore, that no universal agreement has as yet been reached regarding the precise geological time and taxonomic place of the emergence of the first real hominids nor of the first true man. These questions have still to be answered in the form of a compromise between the views of the many different specialists. At present the most convincing assumption appears to be that a clear and decisive separation of Hominidae had taken place at some time in the earlier part of the Miocene period, when there had existed a remarkably flourishing and widely distributed group of early hominoids with as yet very limited specialisations. The pedigree in figure 79 shows in diagrammatic form this interpretation of the evolutionary relations between the higher primates in general and man in particular.

Except for some fragmentary remains of a hominoid primate, called *Ramapithecus*, with comparatively short jaws and certain dental peculiarities, known from India, East Africa and possibly Europe and China, no other fossils have so far been found, which might reasonably be claimed to fit near the hominid line of ascent before the Pleistocene period. From the first half of the latter period in South and East Africa have come the very fortunate and startling discoveries of unusually many and extensive remains of fossil teeth, skulls and other skeletal parts, belonging to a clearly hominid genus, known as *Australopithecus* and containing at least two somewhat different species. The great significance of these fossils for the appreciation of human evolution consists first of all in the fact that the construction of their pelvis is so man-like that it leaves very little doubt of their having already acquired the erect posture and bipedal locomotion. This they had combined with a brain of very moderate size and a dentition containing small front teeth, including canines, but large grinding teeth. Judging by the great mass and variety of associated palaeontological and archaeological finds, the australopithecids had lived in savanna-like regions in which they apparently subsisted to a considerable extent by hunting smaller animals with the aid of crude weapons and tools of stone, bone, horn and teeth. In their general appearance and behaviour these close relatives of early Pleistocene man can hardly have been very different from the latter, though their faces were certainly more ape-like, as can be concluded from the detailed construction of their skulls. The smaller type of *Australopithecus*, which was about the size of a male chimpanzee and most likely had a face resembling it, probably still had that simian look of

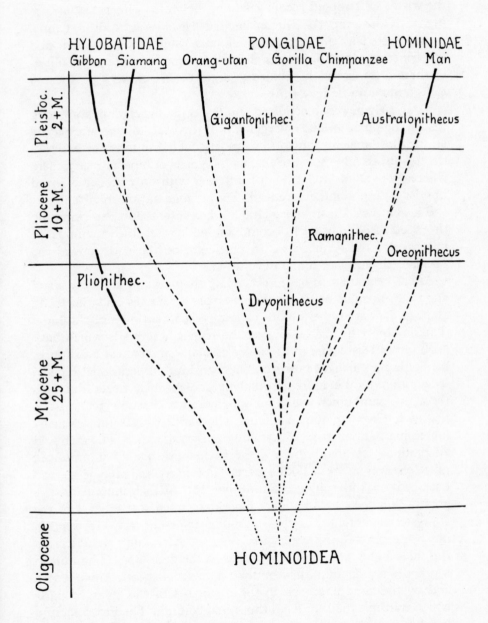

Figure 79 Diagrammatic family-tree of the Hominoidae. M = millions of years.

251

the writer's attempted reconstruction, shown in plate 15. Contrasting with this, the bust of an ancient *Homo erectus* from China, reconstructed in the same manner on a well-preserved skull and illustrated in plate 16, reflects already the emerging human qualities of these creatures with decidedly larger brains, more sophisticated tools and a knowledge of fire.

That early man had acquired and perfected his erect posture long before the final, unique enlargement of the brain is well attested by all the most ancient human remains discovered so far. The bones of the legs invariably show the characters typical of bipedal recent man, whereas the skulls are still very different with only moderate-sized brain cavities, sloping foreheads above huge supraorbital shelves, and heavy jaws. It can be concluded, therefore, that it was primarily the specialised mode of locomotion which set the hominids in general and the genus *Homo* in particular on the path toward the ultimate advances of human evolution.

As has been pointed out in preceding chapters, the ability to stand and walk upright exists in a great many primates and the trunk alone is commonly held more or less erect in sitting and climbing. This is especially the case in all the recent hominoids which have acquired at least some of the more essential anatomical conditions, facilitating an upright posture. The transition of occasional bipedal locomotion to the more permanent one of man never did pass through a semi-erect stage, but was merely a chronological matter, just as in growing human infants who learn to walk for gradually increasing periods in an always fully upright manner. The manifold adaptations for man's highly specialised posture have produced great changes in most bodily parts. The head has become so well balanced on the spinal column that the nuchal musculature is greatly reduced, the shoulders have descended to the level of the suprasternal notch, and the neck appears comparatively slender, not being covered by huge muscles in the back, high shoulders on the sides and a large face in front, as in the great apes. The human hip bones are distinguished by their extremely short relative length and by the close approach to the hip joints of the enlarged parts which transmit the load from the spinal column. The broad sacrum has become bent sharply back, forming a far more marked promontory at the lumbo-sacral border than develops in any of the apes during postnatal life. The lower extremities of man reach an unusual length rather late in growth, when they exceed the length of the

252

arms, in contrast to the other hominoids, in which this relationship is reversed. With the feet of terrestrial man no longer required for grasping, the first toes have lost most of their former typically simian movability and the phalangeal parts of the other toes have largely degenerated through their unique reduction in length. In the course of his long separate evolution man has gained many further more or less localised specialisations of his body, as have been discussed in previous chapters. Most of the purely quantitative human distinctions, such as those regarding the relative size or position of particular features, reach their full development only during postnatal life, when they often become simply more extreme than in the pongids. The striking reduction of man's coat of hair represents merely the maximum advance of a trend which has also affected the great apes with their comparatively sparse hair density. The long list of characters defining man contains also items regarding which he has remained more conservative than the great apes, as for instance the phylogenetic changes in the regional numbers of vertebrae and the avoidance of the more extreme specialisations in the limbs of the latter, or in their outer ears, throat pouches, etc.

Ontogenetically man has become distinguished not only by a great variety of retardations as well as accelerations in developmental processes, but also by the extreme prolongation of his postnatal growth period and entire life span. With the increased duration of immaturity the helpless human newborn require intense maternal care for an exceptionally long time during which they benefit from correspondingly extended opportunities for learning in close association with their elders. The progressing use and later manufacture of tools for defensive and constructive purposes had become facilitated chiefly by the vastly perfected cerebral control of the hands which have otherwise undergone comparatively little specialisation. The elaboration of man's central nervous system has also played a far more vital role in the development of language than have any of the minor changes in our speech apparatus. The admirable ability to employ words as symbols in the exchange of information and announcement of intentions must have evolved in closest connection with the social life of early man. The latter had to become adapted to the new conditions for survival of a bipedal, non-arboreal primate, lacking great speed and powerful canines for its defence and needing effective means for communication for successful hunting through co-operation. When carnivorous man no longer feared fire, but had learned to use it in the preparation of food and for warmth in addition to hides or other clothing, he was ready to venture into colder climates and to invade vast new regions, unsuited for the more limited adaptability of other primates. The initially slow and finally rapid cultural progress of mankind has resulted in its unique expansion over five continents and even into arctic zones and is mainly responsible for the alarming increase in the population of recent man.

Man's widespread deforestation for gaining new land in the competition for food is ruthlessly restricting the natural habitats of other primates. The latter are being used for modern research in staggering numbers. It appears, therefore, that the most successful of all primates – man – is seriously interfering with the survival of the remaining nonhuman primates which have been described in this book.

Bibliography

THIS bibliography contains first of all the most essential references to publications used in preparing the text and figures of each chapter. In addition there have been listed the titles of some of the latest and most useful books, from which the interested reader can gain a great deal more detailed information on the past and present life of primates. Many of these books deal with several or even most of the major topics in the present volume and might properly have been repeated in the lists of literature for each chapter to which they appertain. To save space, however, such titles are generally given only once, but are marked with an asterisk, as are also the modern collections of reprinted articles of outstanding significance and those of papers contributed to symposia on general problems, such as primate behaviour, locomotion, etc. The names and subjects of all the many different authors in the latter volumes have here not been listed separately, so that this bibliography would not become unduly long.

Preface and Chapter 1, Historical Introduction

CAMPER, P. (1779) Account of the organs of speech of the orang outang. *Philos. Transact.*, London, **69**: 139–150

CARPENTER, C.R. (1964) *Naturalistic behavior of nonhuman primates.* Pennsylvania State University Press

*CLARK, W.E. Le Gros (1962) *The antecedents of man. An introduction to the evolution of the primates.* 2nd ed. Edinburgh, at the University Press

*DE VORE, I. (Ed.) (1965) *Primate behavior. Field studies of monkeys and apes.* Holt, Rinehart & Winston, New York

DU CHAILLU, P.B. (1861) *Explorations and adventures in equatorial Africa.* Harper Bros., New York

ELLIOT, D.G. (1913) *A review of the primates.* (3 vols.) Monograph No. 1, Amer. Mus. Nat. Hist., New York

GARNER, R.L. (1900) *Apes and monkeys, their life and language.* Athenaeum Press, London

*GAVAN, J.A. (Ed.) (1955) *The non-human primates and human evolution.* Wayne University Press, Detroit

*HILL, W.C.O. (1953–66) *Primates. Comparative anatomy and taxonomy.* Vols. I–VI (to be continued). Edinburgh, at the University Press

*HOFER, H., SCHULTZ, A.H. and STARCK, D. (Ed.) (1956–69) *Primatologia. Handbook of Primatology.* Vols. I and III and many chapters of other volumes, issued separately in advance (to be continued). Karger, Basel

HOOTON, E. (1942) *Man's poor relations.* Doubleday, Doran & Co., Garden City, New York

HUXLEY, T.H. (1863) *Evidence as to man's place in nature.* Williams and Norgate, London

JANSON, H.W. (1952) *Apes and ape lore in the middle ages and the renaissance.* Studies of the Warburg Inst. Vol. 20, London

LINNAEUS, C. (1758) *Systema naturae per regna tria naturae.* T. I. Editio decima. Holmiae, Laurentii Salvii

McDERMOTT, W.C. (1938) *The ape in antiquity.* Johns Hopkins Press, Baltimore

MORRIS, R. and MORRIS, D. (1966) *Men and apes.* Hutchinson & Co., London

*RUCH, T.C. (1941) *Bibliographia primatologica.* Part I. Thomas, Springfield, Ill.

SCHALLER, G.B. (1963) *The mountain gorilla; ecology and behavior.* University of Chicago Press

TYSON, E. (1699) *Orang-Outang, sive* Homo Sylvestris *or, the anatomy of a pygmie compared with that of a Monkey, an Ape, and a Man. To which is added, A philological essay concerning the Pygmies, the Cynocephali, the Satyrs, and Sphinges of the Ancients. Wherein it will appear that they are all either Apes or Monkeys, and not Men, as formerly pretended.* Thomas Bennet at the Half-Moon in St Paul's Church-yard, London

*YERKES, R.M. and YERKES, A.W. (1929) *The great apes. A study of anthropoid life.* Yale University Press, New Haven

ZUCKERMAN, S. (1932) *The social life of monkeys and apes.* Kegan, Paul & Co., London

Chapter 2, Introductory Survey of Primates

CLARK, W.E.LE GROS (1949) *History of the Primates.* British Museum (Nat. Hist.), London

FIEDLER, W. (1956) Uebersicht über das System der Primates. In: *Primatologia* 1: 1–267. Karger, Basel

HILL, W.C.O. (1953–66) *Primates. Comparative anatomy and taxonomy.* Vols. I–VI. Edinburgh, at the University Press

JOLLY, C.J. (1966) Introduction to the Cercopithecoidea, with notes on their use as laboratory animals, *Symposia, Zoolog. Soc.,* London, No. 17: 427–57

KUHN, H.-J. (1967) Zur Systematik der Cercopithecidae. In: *Neue Ergebnisse der Primatologie* (D. Starck, R. Schneider u. H.-J. Kuhn, Ed.) 25–46. Fischer, Stuttgart

NAPIER, J.R. and NAPIER, P.H. (1967) *A handbook of living primates.* Academic Press, London

REYNOLDS, V. (1968) *The apes.* Cassell & Co., London

SCHULTZ, A.H. (1933) Observations on the growth, classification and evolutionary specialization of gibbons and siamangs. *Human Biol.,* **5:** 212–55, 385–428

SCHULTZ, A.H. (1954) Bemerkungen zur Variabilität und Systematik der Schimpansen. *Saugetierkundl. Mitteil.,* **2:** 159–63

SIMPSON, G.G. (1945) The principles of classification and a classification of mammals. *Bull. Amer. Mus. Nat. Hist.* **85:** I–XVI, 1–350

SIMPSON, G.G. (1962) Primate taxonomy and recent studies of nonhuman primates. *Ann. New York Acad. Sciences,* **102,** 2: 497–514

VALLOIS, H. (1955) Ordre des primates. In: *Traité de Zoologie* (P.P. Grassé Ed.) T. **XVII:** 1854–2206. Masson & Cie, Paris

*WASHBURN, S.L. (Edit.) (1963) Classification and human evolution. Viking Fund Publ. in *Anthropol.,* 37

Chapter 3, Origin and Distribution of Primates in Time and Space

GENET-VARCIN, E. (1963) *Les singes actuels et fossiles.* Boubée & Cie., Paris

McKENNA, M.C. (1966) Paleontology and the origin of the primates. *Folia primatologica,* **4:** 1–25

PATTERSON, B. (1955) The geologic history of non-hominid primates in the Old World. In: *The non-human Primates and Human Evolution* (J.A. Gavan Ed.), 13–31. Wayne University Press, Detroit

PIVETEAU, J. (1957) *Primates. Palaeontologie humaine. Traité de Palaeontologie,* T. VII. Masson & Cie., Paris

REMANE, A. (1956) Palaeontologie und Evolution der Primaten. In: *Primatologia* **1:** 267–378. Karger, Basel

SIMONS, E.L. (1963) A critical reappraisal of Tertiary primates. In: *Evolutionary and Genetic Biology of Primates* (J. Buettner-Janusch Ed.) **1:** 65–129. Academic Press, New York

SIMONS, E.L. and PILBEAM, D.R. (1965) Preliminary revision of the Dryopithecinae (Pongidae, Anthropoidea). *Folia primatologica,* **3:** 81–152

THENIUS, E. UND HOFER, H. (1960) *Stammesgeschichte der Säugetiere.* Springer, Berlin

WILSON, J.A. (1966) A new primate from the earliest Oligocene, West Texas, preliminary report. *Folia primatologica,* **4:** 227–248

Chapter 4, Factors Determining Distribution

GUILLOUD, N.B. AND FITZ-GERALD, F.L. (1967) Spontaneous hyperthermia in the gorilla. *Folia primatologica,* **6:** 177–9

HALL, K.R.L. (1966) Distribution and adaptations of baboons. *Symposia Zoolog. Soc.* London, No. **17:** 49–73

KERN, J.A. (1964) Observations on the habits of the proboscis monkey, *Nasalis larvatus* (Wurmb), made in Brunei Bay area, Borneo. *Zoologica* (New York Zoolog. Soc.), **49:** 183–192

*KRUMBIEGEL, I. (1954) *Biologie der Säugetiere.* 2 vols. Agis Verlag Krefeld

KUMMER, H. AND KURT, F. (1963) Social units of a free-living population of hamadryas baboons. *Folia primatologica,* **1:** 4–19

*MORRIS, D. AND JARVIS, C. (Ed.) (1960) *The international zoo yearbook,* I. Zoolog. Soc., London

TAPPEN, N.C. (1960) Problems of distribution and adaptation of the African monkeys. *Current Anthropology,* **1:** 91–120

WADA, K. (1964) Some observations on the life of monkeys in a snowy district of Japan. *Physiol. and Ecology,* **12:** 151–74

Chapter 5, Posture and Locomotion

AVIS, V. (1962) Brachiation: The crucial issue for man's ancestry. *Southwestern J. Anthropol.,* **18:** 119–48

BISHOP, A. (1964) Use of the hand in lower primates. In: *Evolutionary and Genetic Biology of Primates* (J.Buettner-Janusch, Ed.) **2:** 133–225. Academic Press, New York

ERIKSON, G.E. (1963) Brachiation in New World monkeys and in anthropoid apes. *Symposia, Zoolog. Soc.,* London, **10:** 135–64

HRDLICKA, A. (1931) *Children who run on all fours: and other animal-like behaviors in the human child.* McGraw-Hill Book Co., New York

*KINZEY, W.G. (Ed.) (1967) Symposium on primate locomotion. *Amer. J. Phys. Anthropol.,* **26:** 115–275

NAPIER, J.R. (1961) Prehensility and opposability in the hands of primates. *Symposia, Zoolog. Soc.,* London, **5:** 115–31

NAPIER, J.R. (1963) Brachiation and brachiators. *Symposia, Zoolog. Soc.,* London, **10:** 183–95

NAPIER, J.R. AND WALKER, A.C. (1967) Vertical clinging and leaping – a newly recognized category of locomotor behaviour of primates. *Folia primatologica,* **6:** 204–19

PRIEMEL, G. (1937) Die platyrrhinen Affen als Bewegungstypen. Unter besonderer Berücksichtigung der Extremformen *Callicebus* und *Ateles.* *Z. f. Morphol. u. Oekol. d. Tiere,* **33:** 1–52

RIESEN, A.H. AND KINDER, E.F. (1952) *Postural development of infant chimpanzees.* Yale University Press, New Haven

SCHULTZ, A.H. (1933) Die Körperproportionen der erwachsenen catarr-hinen Primaten, mit spezieller Berücksichtigung der Menschenaffen. *Anthropol. Anz.*, **10**: 154–85

SCHULTZ, A.H. (1954) Studien über die Wirbelzahlen und Körper-proportionen von Halbaffen. Vierteljahrsschr. *Naturforsch*. Ges. Zürich, **99**: 39–75

SCHULTZ, A.H. (1961) Vertebral column and thorax. In: *Primatologia*, **4**, Liefer. 5: 1–66. Karger, Basel

SCHULTZ, A.H. (1968) Form und Funktion der Primatenhände. In: *Handgebrauch und Verstandigung bei Affen und Fruhmenschen* (B. Rensch, Ed.): 9–25. Huber, Bern

SPRANKEL, H. (1965) Untersuchungen an Tarsius. *Folia primatologica* **3**: 153–88

TUTTLE, R.H. (1967) Knuckle-walking and the evolution of the hominoid hands. *Amer. J. Phys. Anthropol.*, **26**: 171–206

Chapter 6, Skeleton

ANKEL, F. (1967) Morphologie von Wirbelsäule und Brustkorb. *Prima-tologia*, **4**, Liefer. 4: 1–120

CHOPRA, S.R.K. (1957) The cranial suture closure in monkeys. *Proc. Zoolog. Soc.*, London, **128**: 67–112

*GREGORY, W.K. (Ed.) (1950) *The anatomy of the gorilla*. The Henry Cushier Raven Memorial Vol., Columbia University Press, New York

HOFER, H. (1965) Die morphologische Analyse des Schädels des Men-schen. In: *Menschliche Abstammungslehre* (G.Heberer, Ed.): 145–226. Fischer, Stuttgart

KNUSSMANN, R. (1967) Humerus, Ulna und Radius der Simiae. *Biblio-theca primatologica*, Fasc. 5. Karger, Basel

RANDALL, F. (1943–44) The skeletal and dental development and variability of the gorilla. *Human Biol.*, **15**: 236–54, 307–37, **16**: 23–76

SCHULTZ, A.H. (1930) The skeleton of the trunk and limbs of higher primates. *Human Biol.*, **2**: 303–438

SCHULTZ, A.H. (1937) Proportions, variability and asymmetries of the long bones of the limbs and the clavicles in man and apes. *Human Biol.*, **9**: 281–328

SCHULTZ, A.H. (1953) The relative thickness of the long bones and the vertebrae in primates. *Amer. J. Phys. Anthropol.* n. s. **11**: 277–312

SCHULTZ, A.H. (1955) The position of the occipital condyles and of the face relative to the skull base in primates. *Amer. J. Phys. Anthropol.*, n. s. **13**: 97–120

SCHULTZ, A.H. (1961) Vertebral column and thorax. In: *Primatologia*, **4**: Liefer. 5: 1–66

SCHULTZ, A.H. (1963) Relations between the lengths of the main parts of the foot skeleton in primates. *Folia primatologica*, **1**: 150–71

VOGEL, C. (1966) Morphologische Studien am Gesichtsschädel catarr-hiner Primaten. *Bibliotheca primatologica*, Fasc. 4. Karger, Basel

Chapter 7, Oral Cavity

FRISCH, J.E. (1965) Trends in the evolution of the hominoid dentition. *Bibliotheca primatologica*, Fasc. 3. Karger, Basel

GREGORY, W.K. (1920–21) The origin and evolution of the human dentition. *J. Dental Research*, **2**: 89–183, 215–83, 357–426, 607–717, **3**: 87–228

JAMES, W.W. (1960) *The jaws and teeth of primates*. Pitman Medical Publishing Co., London

REMANE, A. (1962) Zähne und Gebiss. In: *Primatologia*, **3**, T. 2: 637–846. Karger, Basel

SCHNEIDER, R. (1958) Vestibulum oris und Morphologie der Mundspei-cheldrüsen. – Zunge und weicher Gaumen, In: *Primatologia*, **3**, T. 1: 5–40, 61–126. Karger, Basel

SCHULTZ, A.H. (1935) Eruption and decay of the permanent teeth in primates. *Amer. J. Phys. Anthropol.*, **19**: 489–581

SCHULTZ, A.H. (1958) Palatine ridges. In: *Primatologia*, **3**, T. 1: 127–38. Karger, Basel

Chapter 8, Skin and Hair

BIEGERT, J. (1961) Volarhaut der Hände und Füsse. In: *Primatologia*, 2, Liefer, 3: 1–326. Karger, Basel

CRUZ LIMA, E. DA (1945) *Mammals of Amazonia*, I. General introduction and primates (with 42 colored plates by the author). Belém do Pará – Rio de Janeiro

EPPLE, G. UND LORENZ, R. (1967) Vorkommen, Morphologie und Funktion der Sternaldrüse bei den Platyrrhini. *Folia primatologica*, **7**: 98–126

MATTHEWS, L.H. (1956) The sexual skin of the gelada baboon (*Theropithecus gelada*). *Transact. Zoolog. Soc.*, London, **28**: 543–53

MILLER, G.S., JR. (1931) *Human hair and primate patterning*. Smithsonian Miscell. Coll., **85**, No. 10: 1–13

MONTAGNA, W. (1962) The skin of lemurs. *Annals, New York Acad. Sciences*, **102**: 190–209

SCHULTZ, A.H. (1931) The density of hair in primates. *Human Biol.*, **3**: 303–21

SCHWALBE, G. (1911) *Ueber die Richtung der Haare bei den Affenembryonen. Studien über Entwickelungsgeschichte der Tiere* (E. Selenka, Ed.) 15. H.: Menschenaffen, 10. Liefer. Kreidel, Wiesbaden

Chapter 9, Brain

CONNOLLY, C. J. (1950) *External morphology of the primate brain.* Thomas, Springfield, Ill.
HOFER, H. (1962) Ueber die Interpretation der ältesten fossilen Primatengehirne. In: *Bibliotheca primatologica,* **1:** 1–31. Karger, Basel
SCHULTZ, A. H. (1941) The relative size of the cranial capacity in primates. *Amer. J. Phys. Anthropol.,* **28:** 273–87
SCHULTZ, A. H. (1965) The cranial capacity and the orbital volume of hominoids according to age and sex. In: *Homenaje a Juan Comas.* Vol. **2:** 337–57. Mexico
STARCK, D. (1965) Die Neencephalisation (Die Evolution zum Menschenhirn). In: *Menschliche Abstammungslehre* (G. Heberer, Ed.) 103–44. Fischer, Stuttgart
STEPHAN, H.(1967) Zur Entwicklungshöhe der Primaten nach Merkmalen des Gehirns. In: *Neue Ergebnisse der Primatologie* (D. Starck, R. Schneider, H.–J. Kuhn, Ed.): 108–19. Fischer, Stuttgart

Chapter 10, Special Sense Organs

LASINSKI, W. (1960) Aeusseres Ohr. In: *Primatologia,* **2,** T. 1, Liefer. 5: 41–74. Karger, Basel
MOYNIHAN, M. (1966) Communication in the titi monkey, *Callicebus. J. Zoology* (London), **150:** 77–127
POLYAK, S. (1957) *The vertebrate visual system.* University of Chicago Press
ROHEN, J.W. (1962) Sehorgan. In: *Primatologia,* **2,** T. 1, Liefer. 6: 1–210. Karger, Basel
SCHULTZ, A. H. (1935) The nasal cartilages in higher primates. *Amer. J. Phys. Anthropol.,* **20:** 205–12
SCHULTZ, A. H. (1940) The size of the orbit and of the eye in primates. *Amer. J. Phys. Anthropol.,* **26:** 389–408
WEN, I. C. (1930) Ontogeny and phylogeny of the nasal cartilages in primates. *Contrib. to Embryol.* Carnegie Inst. Washington Publ. 414: 109–34
WERNER, C. F. (1960) Mittel- und Innenohr. In: *Primatologia,* **2,** T. 1, Liefer. 5: 1–40

Chapter 11, Growth and Development

GAVAN, J. (1953) Growth and development of the chimpanzee; a longitudinal and comparative study. *Human Biol.,* **25:** 93–143
GAVAN, J.A. AND SWINDLER, D.R. (1966) Growth rates and phylogeny in primates. *Amer. J. Phys. Anthropol.,* n. s. **24:** 181–90
NISSEN, H.W. AND RIESEN, A.H. (1949) Onset of ossification in the

epiphyses and short bones of the extremities in chimpanzees. *Growth*, **13**: 45–70

RIOPELLE, A. J. (1963) Growth and behavioral changes in chimpanzees. *Z. f. Morphol. u. Anthropol.*, **53**: 53–61

SCHULTZ, A. H. (1926) Fetal growth of man and other primates. *Quart. Rev. Biol.*, **1**: 465–521

SCHULTZ, A. H. (1933) Chimpanzee fetuses. *Amer. J. Phys. Anthropol.* **18**: 61–79

SCHULTZ, A. H. (1940) Growth and development of the chimpanzee. *Contrib. to Embryol.* 28, Carnegie Inst. Washington Publ. 518: 1–63

SCHULTZ, A. H. (1941) Growth and development of the orang-utan. *Contrib. to Embryol.*, 29, Carnegie Inst. Washington Publ. 525: 57–110

SCHULTZ, A. H. (1942) Growth and development of the proboscis monkey. *Bull. Mus. Compar. Zoology*, Harvard College, **89**: 279–314

SCHULTZ, A. H. (1944) Age changes and variability in gibbons. *Amer. J. Phys. Anthropol.*, n. s. **2**: 1–129

SCHULTZ, A. H. (1956) Postembryonic age changes. In: *Primatologia*, **1**: 887–964. Karger, Basel

WAGENEN, G. VAN AND ASLING, C. W. (1964) Ossification in the fetal monkey (*Macaca mulatta*). *Amer. J. Anat.*, **114**: 107–132

Chapter 12, Reproduction, Morbidity and Mortality

ALTMANN, S. A. (1962) A field study of the sociobiology of rhesus monkeys, *Macaca mulatta*. *Ann. New York Acad. Sciences*, **102**: 338–435

ASDELL, S. A. (1946) *Patterns of mammalian reproduction*. Comstock Publish. Co., Ithaca, N. Y.

BOOTH, C. (1962) Some observations on behavior of Cercopithecus monkeys. *Ann. New York Acad. Sciences*, **102**: 477–87

BOURLIÈRE, F. ET PETTER-ROUSSEAUX, A. (1966) Existence probable d'un rythme metabolique saisonnier chez les Cheirogaleinae (Lemuroidea). *Folia primatologica*, **4**: 249–56

BUTLER, H. (1967) The oestrus cycle of the Senegal bushbaby (*Galago senegalensis senegalensis*) in the Sudan. *J. Zoology* (London), **151**: 143–62

CARPENTER, C. R. (1940) A field study in Siam of the behavior and social relations of the gibbon (*Hylobates lar*). *Compar. Psychol. Monogr.*, **16**, No. 5

COLYER, F. (1936) *Variations and diseases of the teeth of animals*. J. Bale, Sons & Danielsson, London

CONAWAY, C. H. AND SADE, D. S. (1965) The seasonal spermatogenic cycle in free ranging rhesus monkeys. *Folia primatologica*, **3**: 1–12

COWGILL, U. M. (1964) Recent variations in the season of birth in Puerto Rico. *Proc. Nat. Acad. Sciences*, **52**: 1149–51

DUCKWORTH, W. L. H. (1912) On the natural repair of fractures, as seen in the skeleton of anthropoid apes. *J. Anat. & Physiol.*, **46**: 81–5

DUNN, F.L. (1966) Patterns of parasitism in primates: Phylogenetic and ecological interpretations, with particular reference to Hominoidea. *Folia primatologica,* **4**: 329–45

*FIENNES, R.N.T-W- (Ed.) (1966) Some recent developments in comparative medicine. *Symposia Zoolog. Soc.* London, **17.** Academic Press, London

FOX, H. (1939) Chronic arthritis in wild mammals. *Transact. Amer. Philos, Soc.*, n. s. **31:** 71–149

GRAY, A.P. (1954) *Mammalian hybrids. A check-list with bibliography.* Technical Communication No. 10, Commonwealth Bureau of Animal Breeding and Genetics. Edinburgh

HARMS, J.W. (1956) Fortpflanzungsbiologie. In: *Primatologia,* **1:** 562–660. Karger, Basel

HILL, W.C.O. (1958) External genitalia. In: *Primatologia,* **2,** T. 1: 630–704

KOFORD, C.B. (1963) Group relations in an island colony of Rhesus monkeys. In: *Primate Social Behavior* (C.H. Southwick, Ed.): 136–52. Van Nostrand Co., Princeton, N. J.

LANCASTER, J.B. AND LEE, R.B. (1965) The annual reproductive cycle in monkeys and apes. In: *Primate Behavior* (I.De Vore, Ed.): 486–513. Holt, Rinehart & Winston, New York

LAPIN, B.A. UND JAKOVLEVA, L.A. (1964) *Vergleichende Pathologie der Affen.* Fischer, Jena

MALINOW, M.R. (1965) Atherosclerosis in subhuman primates. *Folia primatologica,* **3:** 277–300

MILLER, G.S., JR. (1931) The primate basis of human sexual behavior. *Quart. Rev. Biol.*, **6:** 379–410

PETTER, J.J. (1962) Recherches sur l'écologie et l'éthologie des lémuriens Malagaches. *Mém. Mus. Nat. d'Hist. Naturelle.* Nouv. Série *A,* Zoologie, **27,** Fasc. 1

PETTER-ROUSSEAUX, A. (1964) Reproductive physiology and behavior of the Lemuroidea. In: *Evolutionary and Genetic Biology of Primates* (J. Buettner-Janusch, Ed.), **2:** 92–132. Academic Press, New York

RIPLEY, S. (1967) Intertroop encounters among Ceylon gray langurs (*Presbytis entellus*). In: *Social Communication among Primates* (S.A. Altmann, Ed.): 237–57. University of Chicago Press

RUCH, T.C. (1959) *Diseases of laboratory primates.* Saunders Co., Philadelphia

*SAUER, R.M. (Ed.) (1960) Care and diseases of the research monkey. *Ann. New York Acad. Sciences,* **85:** 735–992

SCHULTZ, A.H. (1935) Eruption and decay of the permanent teeth in primates. *Amer. J. Phys. Anthropol.*, **19:** 489–581

SCHULTZ, A.H. (1938) The relative weight of the testes in primates. *Anatomical Rec.*, **72**: 387–94

SCHULTZ, A.H. (1948) The number of young at a birth and the number of nipples in primates. *Amer. J. Phys. Anthropol.*, n. s. **6**: 1–24

SCHULTZ, A.H. (1956) The occurrence and frequency of pathological and teratological conditions and of twinning among nonhuman primates. In: *Primatologia*, **1**: 965–1014. Karger, Basel

SCHULTZ, A.H. (1958) Acrocephalo-oligodactylism in a wild chimpanzee. *J. Anat.*, **92**: 568–79

SCHULTZ, A.H. (1960) Age changes and variability in the skulls and teeth of the Central American monkeys *Alouatta, Cebus* and *Ateles*. *Proc. Zoolog. Soc.*, London, **133**: 337–90

ZUCKERMAN, S. (1957) The human breeding season. *The New Scientist*, April, 1957

Chapter 13, Sexual and Intraspecific Differences

ASHTON, E.H. AND ZUCKERMAN, S. (1950) The influence of geographic isolation on the skull of the green monkey (*Cercopithecus aethiops sabaeus*) I. A comparison between the teeth of the St. Kitts and the African green monkey. *Proc. Roy. Soc. B*, **137**: 212–38

ASHTON, E.H. (1960) The influence of geographic isolation on the skull of the green monkey (*Cercopithecus aethiops sabeus*) V. The degree and pattern of differentiation in the cranial dimensions of the St. Kitts green monkey. *Proc. Roy. Soc. B*, **151**: 563–83

BERNSTEIN, I.S. (1966) Naturally occurring primate hybrid. *Science*, **154**: 1559–60

HILL, W.C.O. (1967) Taxonomy of the baboon. In: *The Baboon in Medical Research* (Vagtborg, Ed.), **2**: 3–11. University of Texas Press, Austin, Tex.

KUHN, H.–J. (1964) Zur Kenntnis von Bau und Funktion des Magens der Schlankaffen (Colobinae). *Folia primatologica*, **2**: 193–221

SCHULTZ, A.H. (1944) Age changes and variability in gibbons. *Amer. J. Phys. Anthropol.*, n. s. **2**: 1–129

SCHULTZ, A.H. (1949) Sex differences in the pelves of primates. *Amer. J. Phys. Anthropol.*, n. s. **7**: 401–24

SCHULTZ, A.H. (1958) Cranial and dental variability in Colobus monkeys. *Proc. Zoolog. Soc.*, London, **130**: 79–105

SCHULTZ, A.H. (1962) Metric age changes and sex differences in primate skulls. *Z. f. Morphol. u. Anthropol.*, **52**: 239–55

SCHULTZ, A.H. (1963) Age changes, sex differences and variability as factors in the classification of primates. In: Classification and Human Evolution (S.L. Washburn, Ed.), Viking Fund Publ. in *Anthropol.*, **37**: 85–115

WETTSTEIN, E.B. (1963) Variabilität, Geschlechtsunterschiede und Altersveränderungen bei *Callithrix jacchus* L. *Morphol. Jahrb.*, **104:** 185–271

Chapter 14, Behaviour

*ALTMANN, S.A. (Ed.) (1967) *Social Communication among primates.* University of Chicago Press

ANDREW, R.J. (1963) Trends apparent in the evolution of vocalization in the Old World monkeys and apes. In: *Symposia Zoolog. Soc.* London, **10:** 89–102

ANDREW, R.J. (1964) The displays of the primates. In: *Evolutionary and Genetic Biology of Primates* (J. Buettner-Janusch, Ed.) **2:** 227–309. Academic Press, New York

BOOTH, C. (1962) Some observations on behavior of Cercopithecus monkeys. *Ann. New York Acad. Sciences,* **102:** 477–87

CARPENTER, C.R. (1934) A field study of the behavior and social relations of howling monkeys. *Compar. Psychol. Monographs,* **10,** No. 2: 1–168

CARPENTER, C.R. (1940) A field study in Siam of the behavior and social relations of the gibbon (*Hylobates lar*). *Compar. Psychol. Monographs* **16,** No. 5: 1–212

CROOK, J.H. (1966) Gelada baboon herd structure and movement. A comparative report, *Symposia Zoolog. Soc.* London, **18:** 237–58

*DE VORE, I. (Ed.) (1965) *Primate behavior. Field studies of monkeys and apes.* Holt, Rinehart & Winston, New York

EISENBERG, J.F. AND KUEHN, R.E. (1966) *The behavior of Ateles geoffroyi and related species.* Smithsonian Miscell. Coll., **151,** No. 8

EPPLE, G. (1967) Vergleichende Untersuchungen über Sexual- und Sozialverhalten der Krallenaffen (Hapalidae). *Folia primatologica,* **7:** 37–65

GOODALL, J. (1963) Feeding behaviour of wild chimpanzees. *Symposia, Zoolog. Soc.,* London, **10:** 39–47

HALL, K.R.L. (1962) Numerical data, maintenance activities and locomotion of the wild chacma baboon, *Papio ursinus. Proc. Zoolog. Soc.,* London, **139:** 181–220

HALL, K.R.L. (1962) The sexual, agonistic and derived social behaviour patterns of the wild chacma baboon, *Papio ursinus. Proc. Zoolog. Soc.,* London, **139:** 283–327

HALL, K.R.L. (1966) Behaviour and ecology of the wild patas monkey, *Erythrocebus patas,* in *Uganda. J. Zoology* (London) **148:** 15–87

HALL, K.R.L. AND GARTLAN, J.S. (1965) Ecology and behaviour of the vervet monkey, *Cercopithecus aethiops,* Lolui Island, Lake Victoria. *Proc. Zoolog. Soc.,* London, **145:** 37–56

HUBER, E. (1931) *Evolution of facial musculature and facial expression.* Johns Hopkins Press, Baltimore

JOLLY, A. (1966) *Lemur behavior.* University of Chicago Press

KLÜVER, H. (1933) *Behavior mechanisms in monkeys.* University of Chicago Press

KÖHLER, W. (1927) *The mentality of apes.* 2nd ed., Harcourt, Brace & Co., New York

KORTLANDT, A. (1967) Experimentation with chimpanzees in the wild. In: *Neue Ergebnisse der Primatologie* (D. Starck, R. Schneider, H.–J. Kuhn, Ed.): 208–24. Fischer, Stuttgart

KUMMER, H. (1968) Social organization of hamadryas baboons. *Bibliotheca primatologica*, No. 6, Karger, Basel

LANG, E. M., SCHENKEL, R. UND SIEGRIST, E. (1965) *Gorilla Mutter und Kind.* Basilius Presse, Basel

MORRIS, D. (1967) *Primate ethology.* Weidenfeld & Nicolson, London

MOYNIHAN, M. (1964) *Some behavior patterns of platyrrhine monkeys. I. The night monkey (Aotus trivirgatus).* Smithsonian Miscell. Coll., **146**, No. 5

PETTER, J. J. (1962) Recherches sur l'écologie et l'éthologie des lémuriens malagaches. *Mém. du Mus. Nat. d'Hist. Nat.* Nouv. Sér., *A.* Zoology, **27**, Fasc. 1

PETTER, J. J. AND PETTER, A. (1967) The aye-aye of Madagascar. In: *Social Communications among Primates* (S. A. Altmann, Ed.): 195–205. University of Chicago Press

POPE, B. L. (1968) Population characteristics. In: Biology of the Howler Monkey (*Alouatta caraya*). (M. R. Malinow, Ed.). *Bibliotheca primatologica*, No. **7**: 13–20

REYNOLDS, V. (1965) *Budongo, a forest and its chimpanzees.* Methuen & Co., London

REYNOLDS, V. (1966) Open groups in hominoid evolution. *Man,* **1**: 441–52

*RHEINGOLD, H. L. (Ed.) (1963) *Maternal behavior in mamals.* J. Wiley & Sons, New York

ROWELL, T. E. (1966) Forest living baboons in Uganda. *J. Zoology* (London), **149**: 344–64

SAUER, E. G. F. (1967) Mother-infant relatior ship in galagos and the oral child-transport among primates. *Folia primatologica.* **7**: 127–49

SCHALLER, G. B. (1963) *The mountain gorilla; ecology and behavior.* University of Chicago Press

SCHENKEL, R. AND SCHENKEL-HULLIGER, L. (1967) On the sociology of free-ranging Colobus. In: *Neue Ergebnisse der Primatologie* (D. Starck, R. Schneider, H.–J. Kuhn, Ed.): 185–94. Fischer, Stuttgart

SCHULTZ, A. H. (1961) Some factors influencing the social life of primates in general and of early man in particular. In: Social Life of Early Man (S. L. Washburn, Ed.), Viking Fund Publ. in *Anthropol.*, **31:** 58–90

*SOUTHWICK, C. H. (Ed.) (1963) *Primate social behavior. Selected Readings.* Van Nostrand Co., Princeton, N. J.

STRUHSAKER, T. T. (1967) Behavior of vervet monkeys and other cercopithecines. *Science,* **156:** 1197–1203

THORINGTON, R. W., JR. (1967) Feeding and activity of *Cebus* and *Saimiri* in a Colombian forest. In: *Neue Ergebnisse der Primatologie* (D. Starck, R. Schneider, H.–J. Kuhn, Ed.): 180–84. Fischer, Stuttgart

ULLRICH, T. (1961) Zur Biologie und Soziologie der Colobus-Affen (*Colobus guereza caudatus,* Thomas 1885). *Zoolog. Garten* (N. F.) **25:** 305–68

VAN HOOFF, J. A. R. A. M. (1962) Facial expressions in higher primates. In: Evolutionary aspects of animal communication. *Symposia, Zoolog. Soc.,* London, **8:** 97–125

WASHBURN, S. L., JAY, P. C. AND LANCASTER, J. B. (1965) Field studies of Old World monkeys and apes. *Science,* **150:** 1541–47

Chapter 15, Evolutionary Trends

BREITINGER, E. (1962) Zur gegenwärtigen Kenntnis der ältesten Hominiden. *Anzeiger d. phil.-hist. Kl.,* Oesterreich. Akad. Wissensch., Jahrg. **1961:** 169–207

BUETTNER-JANUSCH, J. (1966) *Origins of man.* Wiley & Sons, New York

CAMPBELL, B. (1966) *Human evolution. An introduction to man's adaptations.* Aldine Publish. Co., Chicago

*CLARK, W. E. LE GROS (1962) *The antecedents of man. An introduction to the evolution of the primates.* 2nd ed. Edinburgh, at the University Press

CLARK, W. E. LE GROS (1967) *Man-apes or ape-man?* Holt, Rinehart and Winston, New York

GREGORY, W. K. (1934) *Man's place among the anthropoids.* Clarendon Press, Oxford

*HEBERER, G. (1961) Die Abstammung des Menschen. In: *Handbuch der Biologie* (L. v. Bertalanffy and F. Gessner, Ed.), **9:** 245–328. Akadem. Verlagsges. Athenaion, Konstanz

*HEBERER, G. (Ed.) (1965) *Menschliche Abstammungslehre.* Fischer, Stuttgart

HOCKETT, C. F. AND ASCHER, R. (1964) The human revolution. *Current Anthropology,* **5:** 135–68

*HOWELLS, W. (Ed.) (1962) *Ideas on human evolution* (Selected essays, 1949–61). Harvard University Press, Cambridge, Mass.

REMANE, A. (1956) Methodische Probleme der Hominidenphylogenie III. Die Phylogenie der Lebensweise und die Entstehung des aufrechten Ganges, *Z. Morphol. u. Anthropol.,* **48:** 28–54

SCHMITT, J. (1968) Immunbiologische Untersuchungen bei Primaten. Ein Beitrag zur Evolution der Blut- und Serumgruppen. *Bibliotheca primatologica*, No. 8. Karger, Basel

SCHULTZ, A.H. (1936) Characters common to higher primates and characters specific for man. *Quart. Rev. Biol.*, **11**: 259–83, 425–55

SCHULTZ, A.H. (1950) The physical distinctions of man. *Proc. Amer. Philosoph. Soc.*, **94**: 428–49

SCHULTZ, A.H. (1968) The recent hominoid primates. In: *Perspectives on Human Evolution* (S.L.Washburn, Ed.): 122–95. Holt, Rinehart & Winston, New York

SIMONS, E.L. (1963) Some fallacies in the study of hominoid phylogeny. *Science*, **141**: 879–89

SIMONS, E.L. (1967) New evidence on the anatomy of the earliest catarrhine primates. In: *Neue Ergebnisse der Primatologie* (D.Starck, R.Schneider, H.–J.Kuhn, Ed.): 15–18. Fischer, Stuttgart

TOBIAS, P.V. (1965) Early man in East Africa. *Science,* **149**: 22–33

*WASHBURN, S.L. (Ed.) (1963) *Classification and human evolution*. Aldine Publish. Co., Chicago

Appendix

The Classification of Recent Primates

ONLY the commonly recognised genera are included and the grouping of these into major categories is chiefly based upon comparative-anatomical evidence which appears at present as most significant.

SUBORDER PROSIMIAE
  INFRAORDER TUPAIIFORMES
    FAMILY TUPAIIDAE
    Genus *Tupaia*   treeshrew
    Genus *Dendrogale*
    Genus *Urogale*
    Genus *Ptilocercus*   pen-tailed treeshrew

  INFRAORDER LEMURIFORMES
    FAMILY LEMURIDAE
    Genus *Lemur*   lemur
    Genus *Hapalemur*   gentle lemur
    Genus *Lepilemur*   weasel lemur
    Genus *Cheirogaleus*   dwarf lemur
    Genus *Microcebus*   mouse lemur

    FAMILY INDRIIDAE
    Genus *Avahi*   avahi
    Genus *Propithecus*   sifaka
    Genus *Indri*   indris

    FAMILY DAUBENTONIIDAE
    Genus *Daubentonia*   aye-aye

  INFRAORDER LORISIFORMES
    FAMILY LORISIDAE
    Genus *Loris*   slender loris
    Genus *Nycticebus*   slow loris
    Genus *Arctocebus*   angwantibo
    Genus *Perodicticus*   potto

FAMILY GALAGIDAE
Genus   *Galago*   bushbaby

INFRAORDER TARSIIFORMES
  FAMILY TARSIIDAE
  Genus   *Tarsius*   tarsier

SUBORDER SIMIAE
INFRAORDER PLATYRRHINA
SUPERFAMILY CEBOIDEA
  FAMILY CALLITHRICIDAE
  Genus   *Callithrix*   marmoset
  Genus   *Leontocebus*   tamarin
  Genus   *Callimico*   Goeldi's tamarin

  FAMILY CEBIDAE
  Genus   *Aotus*   night monkey
  Genus   *Callicebus*   titi monkey
  Genus   *Pithecia*   saki
  Genus   *Chiropotes*   black, etc. saki
  Genus   *Cacajao*   uakarí
  Genus   *Alouatta*   howler monkey
  Genus   *Saimiri*   squirrel monkey
  Genus   *Cebus*   capuchin monkey
  Genus   *Lagothrix*   woolly monkey
  Genus   *Ateles*   spider monkey

INFRAORDER CATARRHINA
SUPERFAMILY CERCOPITHECOIDEA
  FAMILY CERCOPITHECIDAE
  Subfamily Cercopithecinae
  Genus   *Macaca*   macaque
  Genus   *Papio*   baboon
  Genus   *Theropithecus*   gelada
  Genus   *Cercocebus*   mangabey
  Genus   *Cercopithecus*   guenon
  Genus   *Erythrocebus*   patas monkey
  Subfamily Colobinae
  Genus   *Presbytis*   langur
  Genus   *Rhinopithecus*   snub-nosed monkey
  Genus   *Simias*   Pageh snub-nosed monkey
  Genus   *Nasalis*   proboscis monkey
  Genus   *Colobus*   guereza

SUPERFAMILY HOMINOIDEA

FAMILY HYLOBATIDAE

| Genus | *Hylobates* | gibbon |
| Genus | *Symphalangus* | siamang |

FAMILY PONGIDAE

Genus	*Pongo*	orang-utan
Genus	*Pan*	chimpanzee
Genus	*Gorilla*	gorilla

FAMILY HOMINIDAE

| Genus | *Homo* | man |

Index

Italicised page references indicate illustrations